Humber College Library
3199 Lakeshore Blvd. West
Toronto, ON   M8V 1K8

IMPRESSUM
IMPRINT

Herausgeber *Editor* iF International Forum Design GmbH, Bahnhofstrasse 8, 30159 Hannover, Germany, www.ifdesign.de, phone +49.511.54224-0, fax +49.511.54224-100, info@ifdesign.de ▪ **iF Geschäftsführung** *Managing Director* Ralph Wiegmann ▪ **iF design awards 2014** Anja Kirschning, Rylana Kossol, Andrea Schmidt, Carmen Wille, Frank Zierenberg ▪ **iF yearbook 2014** Korinna Braun, Frauke Riechelmann ▪ **iF Team** Dirk Bartelsmeier, Gabriele Bertemann, Sandra Fischer, Heike Meier, Birgit Kölsch, Andrea Schewior, Rainer Schwarz, Annegret Wulf-Pippig ▪ **iF Branch Office Hamburg** Silke Hartung, Hans Pflueg, Nicole Wolf, Hongkongstrasse 6, 20457 Hamburg, Germany ▪ **iF Branch Office Munich** Louisa Erbguth, Petra Nordmeier, Schleißheimer Straße 4, 80333 Munich, Germany ▪ **iF Branch Office Taiwan** Sean C.K. Lee, Tobie Lee, Kimberly Liu, Joan Wu, 3F., No. 133, Guangfu S. Rd., Xinyi Dist., Taipei 110, Taiwan ▪ **iF Representative Office Brasil** Juliana Buso, Centro de Design Paraná, Av. Comendador Franco, 1341 - Jardim Botânico, CEP: 80215-090 Curitiba - Paraná, Brasil ▪ **iF Representative Office Korea** Ji Hwan Lee, Designhouse Inc., Taekwang Building, 162-1 Jangchung-Dong 2-GA, Jung-Gu, Seoul 100-855, Korea ▪ **iF Representative Office Poland** Agnieszka Pagels, Äußere Oybiner Str. 14-16, 02763 Zittau, Germany ▪ **iF Representative Office Turkey** Sinem Kocayas, dDf // dream dream design factory, Eski Şapka Fabrikası, Kumbarahane Cad. No: 22, Haskoy, İstanbul, Turkey ▪ **iF Representative Office Japan** Akiyo Takada, Nahoko Yamashita, #1304 Shibuya Property Tokyu Bldg., 1-32-12, Higashi, Shibuya-ku, Tokyo 150-0011, Japan ▪ **iF Press Office** Claudia Neumann Communication GmbH, Kevin MacArthur, Helena Broustin, Eigelstein 103-113, 50668 Köln, Germany, phone +49.221.9139490, fax +49.221.91394919, iF@neumann-luz.de ▪ **iF DESIGN MEDIA GmbH** Ramona Rockel, phone +49.175.1806786, ramona.rockel@ifdesign.de ▪ **Textredaktion** *Copy Editing* Kristina Irmler, Großburgwedel, Germany; Dr. Tuuli Tietze, Winsen, Germany ▪ **Corporate Design** helke brandt communication, Hannover, Germany ▪ **Übersetzung** *Translation* Lennon.de Language Services, Münster, Germany ▪ **Fotografie Jury** *Jury Photography* Roman Thomas, Celle, Germany; Bernd Schönberger, Berlin, Germany ▪ **Satz und Lithographie** *Typesetting and Lithograph* oeding print GmbH, Braunschweig, Germany ▪ **Druck** *Print* oeding print GmbH, Braunschweig, Germany

Bibliographic information published by the Deutsche Nationalbibliothek. The Deutsche Nationalbibliothek lists this publication in the Deutsche Nationalbibliografie;detailed bibliografic data are available on the internet at http://dnb.d-nb.de This work is subject to copyright. All rights are reserved, whether the whole or part of the material is concerned, specially the rights of translation, reprinting, re-use of illustrations, recitation, broadcasting, reproduction on micofilms or in other ways, and storage in data bases. For any kind of use, permission of the copyright owner must be obtained.

Distribution Partner Prestel Verlag, Munich, A member of Verlagsgruppe Random House GmbH

Prestel Verlag Neumarkter Strasse 28, 81673 Munich, phone +49.89.4136-0, fax +49.89.4136- 2335, www.prestel.de
Prestel Publishing Ltd. 14-17 Wells Street, London W1T 3PD, phone +44.20.7323-5004, fax +44.20.7323-0271, www.prestel.com
Prestel Publishing 900 Broadway, Suite 603, New York, NY 10003, phone +1.212.995-2720, fax +1.212.995-2733, www.prestel.com

Prestel books are available worldwide. Please contact your nearest bookseller or one of the above addresses for information concerning your local distributor. The Library of Congress Cataloguing-in-Publication data is available. British Library Cataloguing-in-Publication Data: a catalogue record for this book is available from the British Library.

© 2014 iF International Forum Design GmbH
Verlag Publisher
iF DESIGN MEDIA GmbH
Bahnhofstrasse 8
30159 Hannover
Germany
www.ifdesign.de

Printed in Germany
ISBN 978-3-7913-4929-9

iF DESIGN AWARDS

product
communication
packaging

2014

INHALT
CONTENT

6	Introduction by Ralph Wiegmann, Managing Director iF
10	Introduction by Fritz Frenkler, Jury Chairman iF design awards 2014
12	Jury iF design awards 2014
18	iF Evaluation Criteria
46	**iF design award winners 2014**
330	Members iF Industrie Forum Design e.V.
338	iF DESIGN TALENTS GmbH
340	Index Manufacturers / Clients
370	Index Designers

TATORT HANNOVER – INVESTIGATING DESIGN

Unser eigener 60. Geburtstag im vergangenen Jahr hat unseren Blick vielleicht in besonderer Weise geschärft: für jene Geschehnisse nämlich, die ein Jubiläum feiern und aufgrund dessen noch einmal die Kontexte, in denen sie seinerzeit entstanden sind, mit dem Heute in Beziehung setzen.

Und 2014 wartet tatsächlich mit einigen besonderen Ereignissen auf, die sich rund jähren. Die bedeutsamsten sind natürlich die großen historischen: Vor 100 Jahren brach der Erste, vor 75 Jahren der Zweite Weltkrieg aus und vor 25 Jahren fiel die Berliner Mauer. Allesamt haben sie das vergangene Jahrhundert maßgeblich geprägt und international zu tiefgreifenden gesellschaftlichen Veränderungen geführt. Der Start des Deutschen Fernsehens im Jahr 1954, ABBAS Erfolg beim Grand Prix 1974, Nelson Mandelas Wahl zum ersten schwarzen Präsidenten Südafrikas 1994 und erst vor zehn Jahren, 2004, der Start von Facebook als internes Netzwerk für die Studenten von Harvard – die „4" am Ende der Jahreszahl steht für viele große und kleine Ereignisse, die auch in unserer Gegenwart Nachhall finden. Und auch die im deutschen Sprachraum erfolgreichste Krimiserie, der „Tatort", wird in diesem Jahr seine 900. Folge zeigen.

Ein spannender Tatort – wenngleich auch ein ganz und gar unblutiger – war im Januar 2014 auch wieder das Messegelände in Hannover. 49 Juroren aus allen Teilen der Welt spürten, Detektiven gleich, nach besonderen Leistungen im Design, nach gelungener oder bedeutender Gestaltung, nach Innovationen und Wegweisendem. Die Motive der Täter, der Ideenentwickler und Formgeber, sind dabei ausschließlich positiv konnotiert: Geht es doch allein darum, den Produkten, den Verpackungen und der Kommunikation eine Gestalt zu geben, die sie einzigartig macht, die Begehrlichkeiten weckt und die den Gebrauch von Produkten erleichtert.

Lassen Sie mich zurückkommen auf die besonderen Ereignisse: Denn mit dem Jahr 2014 verbindet sich auch die Fußballweltmeisterschaft in Brasilien. Für den Designbereich ist das wie ein Fanal, steht Brasilien doch konkret und exemplarisch für den Aufbruch von großen und kleineren aufstrebenden Ländern in die Zukunft. Ein deutlicher Beleg hierfür sind u.a. die seit Jahren steigenden Einreichungen aus jenen ehemaligen Schwellenländern zu den iF design awards – und tatsächlich konnten brasilianische Unternehmen allein im iF communication design award diesmal zwei iF gold awards erringen, an einem weiteren im iF product design award war ein brasilianisches Designbüro beteiligt.

Unsere Jury war sehr angetan von der Qualität der Einreichungen aus Brasilien, aber auch von solchen aus anderen aufstrebenden Ländern beeindruckt – und auch die Produkte aus dem asiatischen Raum halten den Kriterien der iF design awards zunehmend stand. Vieles hat sich getan in China, Korea und Taiwan, auch und gerade im Design mit seiner so wichtigen Rolle als Taktgeber, Vermittler zwischen Industrie und Verbraucher, zwischen Moderne und Tradition. Insbesondere bei den chinesischen Einreichungen stellte unsere Jury fest, dass die Unternehmen und Designstudios sich mit neuen Ideen positioniert und zu eigenen Innovationen gefunden haben. Zwei Produkte aus China und eines aus Taiwan haben in diesem Jahr einen iF gold award gewonnen.

Was kann, abschließend, zu iF generell berichtet werden? Hierfür möchte ich noch einmal auf unseren eigenen Jahrestag, den 60. Geburtstag im letzten Jahr, zurückkommen. Unser Jubiläum hat der Öffentlichkeit und auch uns selbst den großen Sprung verdeutlicht, den iF vor allem im letzten Jahrzehnt gemacht hat. So hat die diesjährige Preisverleihung bereits zum dritten Mal in Folge in der großartigen BMW-Welt in München stattgefunden – wieder konnten wir über 2.000 hochkarätige Gäste begrüßen. Die Verlegung der glamourösen Preisverleihung nach München war ein wichtiger und richtiger Schritt, um die gewachsene Bedeutung der iF design awards zu dokumentieren.

Dann ist im Herzen von Hamburgs HafenCity die neue iF Flagship Ausstellung entstanden – Sie alle sind herzlich eingeladen, in den neuen Elbarkaden eine spannungsvolle Inszenierung von Design zu erleben. Auch dieser Schritt wird iF neue Sichtbarkeit bei neuen Besucher- und Zielgruppen bescheren. Und wem die Zeit oder die Gelegenheit zu einem solchen Ausflug fehlt oder wer sich gerne mittels Smartphone und Tablet informiert, dem sei unsere neue eigene iF design App empfohlen, die erstmalig die digitale Dokumentation der iF design awards bietet – auch dies ein zukunftsweisender Schritt.

In eigener Sache möchte ich an dieser Stelle noch kurz erwähnen, dass ich – gemeinsam mit wunderbaren Kollegen aus der ganzen Welt – seit Januar 2014 Mitglied im Board von Icsid bin, dem Weltdachverband aller Designinstitutionen. Als Geschäftsführer unserer umfassend international aufgestellten und agierenden iF International Forum Design GmbH mit ihren vielen guten Kontakten werde ich diese Rolle natürlich dazu nutzen, im Sinne einer zukunftsorientierten Vorstellung von Design weltweit zu vernetzen, zu verknüpfen und wichtige Themen voranzubringen.

Im Namen unserer iF Familie, die inzwischen aus der iF International Forum Design GmbH, der iF DESIGN TALENTS GmbH, der iF DESIGN MEDIA GmbH, der iF UNIVERSAL + SERVICE GmbH und der iF DESIGN ASIA Ltd. sowie dem iF Industrie Forum Design e.V. besteht, wünschen wir Ihnen viel Freude und Inspiration bei der Lektüre des aktuellen Jahrbuchs!

**HERZLICH,
IHR RALPH WIEGMANN
iF GESCHÄFTSFÜHRER**

CRIME SCENE HANNOVER: INVESTIGATING DESIGN

Our own 60th anniversary last year might have sharpened our awareness in a special way: an awareness for dates that commemorate significant anniversaries and, therefore, for the way in which those events relate the original context to the present day.

The year 2014 will, indeed, bring us a number of anniversaries, the most important of which are, of course, the big historical ones: one hundred years ago, the First World War broke out, seventy-five years ago the Second World War started and twenty-five years ago the Berlin Wall came down. All these events have shaped the last century in significant ways and have, internationally, brought about far-reaching social change. The beginning of television broadcasting in Germany in 1954, ABBA's success at the 1974 Eurovision Song Contest, Nelson Mandela's election to become the first black president of South Africa in 1994 and, only ten years ago in 2004, the launch of Facebook as an internal network for Harvard students: the '4' at the end of the year is representative of many events, both large and small, that also resonate in our present time. And, last but not least, one of the most successful crime series in the German-speaking world, "Tatort", will screen its 900th episode this year.

Another suspenseful crime scene, albeit thankfully free of bloodstains, was Hannover´s exhibition centre, where forty-nine jurors from around the world met in January 2014 to sniff out, like detectives, the best design products, particularly successful or significant designs and innovative and pioneering ideas. Here, the motifs of the perpetrators, of the idea developers and formgivers are exclusively positively connoted: the only focus is on giving products, packaging and communication a form that makes them unique, that evokes desire and facilitates use.

But let me get back to those special events, because the year 2014 is also the year of the Football World Cup in Brazil. For design, this is a watershed, since Brazil represents and is a concrete example of the departure of small and large emerging economies towards the future. Clear evidence of this development is, among other things, to be found in the consistently increasing numbers of entries from those countries in the iF design awards.

In fact, Brazilian companies have been successful this time in securing two gold awards in the communication category alone and another gold award in the product category has been given to an entry that was co-designed by a Brazilian design studio. Our jury was very impressed with the quality of entries from Brazil and from other emerging countries. Products from Asian countries have also increasingly been passing the criteria for the iF design award.

Many things have changed in China, Korea and Taiwan, especially in design, with its important role as a pacemaker and mediator between industry and consumers and between modernity and tradition. Particularly in relation to the Chinese entries, our jury noted that companies and design studios have positioned themselves with new ideas and own innovations. Two Chinese products and one Taiwanese entry have received an iF gold award this year.

So, to conclude, what can I tell you about iF in general? To answer this question, I would like to return again to our own anniversary last year, which made it clear once again, to the public and to ourselves, what an enormous leap iF has made over the last decade: the last award ceremony, for instance, took place for the third time at BMW World in Munich, enjoying a huge audience of more than 2,000 high-profile guests, something we hope will be repeated this year. Relocating the glamorous award ceremony to Munich was an important and necessary step to mirror the growing significance of the iF design awards.

Furthermore, the new iF flagship exhibition opened in the heart of Hamburg's HafenCity and you are warmly invited to visit the new Elb Arcade building and enjoy this exciting showcase of design. This step, too, will entail increased visibility and visitor numbers, as well as new target groups for iF. And those who do not have the time or opportunity to come to Hamburg, as well as those who like to get their information via smartphones and tablets, I would like to point to our new app: since last year, the digital documentation of the iF design awards has been completed, another future-oriented step.

As for myself, I would briefly like to let you know that, in January 2014, I joined the ICSID Board, the worldwide umbrella association of all design institutions, where I have the pleasure of working together with wonderful colleagues from around the world. As the CEO of our very internationally-oriented iF International Forum Design GmbH, I will, of course, use this role and the many contacts it entails to further worldwide networking and to push important themes in the sense of a future-oriented idea of design.

In the name of the entire iF family, which today comprises the iF International Forum Design GmbH, the iF DESIGN TALENTS GmbH, the iF DESIGN MEDIA GmbH and the iF DESIGN ASIA Ltd., as well as the iF Industrie Forum Design e.V., I wish you an enjoying and inspiring read with this, our latest yearbook.

**WARMEST REGARDS,
RALPH WIEGMANN**
iF MANAGING DIRECTOR

THE MUNICH MANIFESTO

Es ist nun wirklich an der Zeit, dass wir uns von dem Wort DESIGN verabschieden. Oder noch schlimmer: von Begriffen wie DESIGNING oder auch ARCHITECTURAL DESIGN.

Der inflationäre Gebrauch des Begriffs DESIGN und seine Verkopplung mit anderen Wörtern werden zunehmend unerträglich, und meistens entsteht nur noch Bedeutungsloses. In der Industrie-Kunst (Industrial Art oder auch Autoren-Design) und auch im Industrie-Entertainment (Industrial Entertainment) wurde der Begriff DESIGN immer schon strapaziert. Auch im Zusammenhang mit den Biofakten und der NSA kommt DESIGN vor. Ich weiß schon lange nicht mehr, wofür diese Wörter alles stehen sollen. Wissen Sie es?

Es macht natürlich auch keinen Sinn, hohe staatliche und auch private Investitionen in eine Berufsausbildung von Designern zu stecken, wenn am Ende kein Mensch weiß, was dabei herauskommt, wohin diese Ausbildung führt.

Gestaltung hat in erster Linie mit Sehen, Denken und Verstehen zu tun und das, was sie bedeutet, muss auch heute noch auf den römischen Architekten, Ingenieur und Architekturtheoretiker Marcus Vitruvius Pollio – Vitruv (1. Jahrhundert vor Christus) – zurückgeführt werden. Die Grundlagen der modernen Gestaltung – vielleicht auch der zweiten Moderne – sind nach Vitruv: „Firmitas", das Ingenieurwesen, „Utilitas", die Wissenschaft, und „Venustas", die Ästhetik. Und diese Grundlagen gelten bis heute.

Wenn dieses verstanden wird und weiterhin Aspekte wie Technik, Ergonomie, Soziologie, Ökologie, Psychologie und Philosophie nach aktueller Erkenntnis berücksichtigt werden, beeinflusst von den bisher bekannten vier unterschiedlichen Dimensionen, dann ergibt dieser Prozess, gekonnt angewendet, gute Gestaltung im Sinne einer Neuen Funktionellen Gestaltung. Dass all dieses natürlich nicht von dem Gestalter allein bewältigt werden kann, ist selbstverständlich und hat einen inter- und multidisziplinären Arbeitsprozess zur Folge, bei dem der Gestalter die Moderation übernimmt.

Und deshalb wurde am 11. Dezember 2013 auf den Vorschlag des Kognitionswissenschaftlers Don Norman hin gemeinsam von Dieter Rams, Herbert Schultes, dem Architekten Thomas Herzog und mir in der Bar dell'Osteria „The Munich Manifesto" beschlossen. Es besagt:
Das Wort DESIGN wird international durch GESTALTUNG ersetzt und Designer werden wieder zu GESTALTERN.

Natürlich ist mir bewusst, wo ich so etwas proklamiere: hier bei iF, bei einer Designinstitution. Aber ich denke, dass iF sehr gut ohne das Wort DESIGN auskommen kann. Jeder, der etwas mit der Gestaltung unserer Mitwelt und Umwelt zu tun hat, kennt iF. Die Marke steht von Beginn an – seit den 1950er Jahren – für gute, verständliche, zweckdienliche, materialgerechte und umweltschonende Gestaltung.
Für eine zeitgemäße und moderne, aber nicht modische Gestaltung – und so unterschied und unterscheiden sich die iF Auszeichnungen grundlegend von vielen anderen Designpreisen.

Auch das Marketing wird so den richtigen Platz in den Unternehmen finden und nicht weiter „Produktentwicklung" betreiben, sondern den Verkauf fördern.

Und so brauchen auch wir, die Gestalter, nicht über Design-Thinking, Experience-Design, Usability-Design, partizipatives Design oder Öko-Design zu sprechen, denn dieses alles war und ist dem Gestaltungsprozess implizit.

FRITZ FRENKLER (GESTALTER)
JURYVORSITZENDER
iF DESIGN AWARDS 2014

THE MUNICH MANIFESTO

The time has come to abandon the word DESIGN as well as DESIGNING and collocations like ARCHITECTURAL DESIGN.

The inflationary trend of using the term DESIGN with its numerous variations has run out of control and is, most of the time, nonsensical, even. Industrial Art (also known as author's design) and Industrial Entertainment always stressed the idea of DESIGN. But today DESIGN is even used when talking about biofacts and the NSA. I just don't understand it anymore. Do you?

Of course, there's no sense in making private and public investments in design education as long as nobody can anticipate the results. Therefore, Don Norman, a US cognitive scientist, declared the Munich Manifesto on 11 December 2013 signed by Dieter Rams, Herbert Schultes, the German architect Thomas Herzog and myself in the Bar dell' Osteria in Munich. The Munich Manifesto states:
DESIGN will internationally be replaced by GESTALTUNG – DESIGNERS will be GESTALTER.

These basic principles, together with the state of our current knowledge of technology, ergonomics, sociology, ecology, psychology and philosophy, and influenced by the four different dimensions, are the fundamentals of the new functional GESTALTUNG.

Of course, these aspects cannot be mastered by an individual GESTALTER alone; rather, this entails an interdisciplinary and multidisciplinary work process in which the GESTALTER takes on the role of moderator.

Even marketing will thus find its rightful place within an organization and will, instead of doing 'product development', promote sales.

Needless to say that I'm conscious of my audience at iF, an institution in the world of design. But I'm confident iF will work without using the word DESIGN. Everyone who is involved in the GESTALTUNG of our everyday world and of our environment, knows iF. From the very beginning, in the 1950s, the brand has been synonymous with good, clear and useful GESTALTUNG, as well as with GESTALTUNG that considers both the use of materials and the environmental impact. By recognizing contemporary and modern GESTALTUNG, the iF awards have always distinguished themselves from other prizes.

First of all, GESTALTUNG is observing, thinking and understanding. Today, GESTALTUNG is still strongly related to the Roman architect, philosopher an engineer Marcus Vitruvius Pollio - Vitruv (1st century BC). He proposed three basic principles for modern GESTALTUNG, including the second modernity: Firmitas (engineering) - Utilitas (science) - Venustas (aesthetics).

And thus we, the GESTALTER, no longer need to talk about design thinking, experience design, usability design, participative design or eco design because all these were, and are, implicit in the process of GESTALTUNG.

FRITZ FRENKLER (GESTALTER)
JURY CHAIRMAN
iF DESIGN AWARDS 2014

JURY iF DESIGN AWARDS 2014
JURY iF DESIGN AWARDS 2014

JURY IMPRESSIONEN
JURY IMPRESSIONS

JURY IMPRESSIONEN
JURY IMPRESSIONS

iF BEWERTUNGSKRITERIEN
iF EVALUATION CRITERIA

Zu den Erfolgsfaktoren der iF design awards gehört ihre Offenheit, nicht nur im Hinblick auf die große Bandbreite an möglichen Wettbewerbsbeiträgen. Auch und gerade die Bewertungskriterien tragen zu dieser Offenheit bei. Sie lassen der Jury Raum für Diskussionen und ermöglichen dadurch fundierte Urteile. Ein Prinzip, das sich bewährt hat – und bereits seit 1975 Gültigkeit beweist. In jenem Jahr stellte Herbert Lindinger die Kriterien auf, an denen sich die iF Wettbewerbe bis heute orientieren.

Mit seiner Definition einer aussagekräftigen, fundierten und unabhängigen Bewertung hat Herbert Lindinger Maßstäbe gesetzt: Die weltweit wichtigsten Designwettbewerbe haben seitdem die von ihm erstellten Bewertungskriterien übernommen. Gleichzeitig ist es für iF als Ausrichter einer der ältesten und renommiertesten Designwettbewerbe Anspruch und Verpflichtung zugleich, diese Kriterien regelmäßig zu überprüfen und an neue Anforderungen und Bedürfnisse anzupassen. Ein Beispiel: Noch vor 20 Jahren spielte „Branding" als Kriterium bei der Bewertung keine Rolle. Heute ist es aus den iF Bewertungskriterien nicht mehr wegzudenken. Dabei folgt iF stets dem Anspruch, vorausschauend zu agieren, statt bloß auf Veränderungen zu reagieren – auch was die Maßstäbe angeht, die wir anlegen, um herausragendes Design zu ermitteln und auszuzeichnen.

One of the key success factors of the iF design awards is their openness, in terms both of the wide range of products exemplifying outstanding design that are eligible to enter for the award and of the evaluation criteria, which embody this spirit of openness perhaps more than anything else. They provide the judging panel with space for wide-ranging discussion and allow them to reach informed decisions. It's a tried-and-trusted principle that has proved valid since 1975, the year Herbert Lindinger laid down the criteria that govern the iF awards to this day.

Herbert Lindinger's definition of pertinent, informed and independent assessment of entries has set standards; since he established the evaluation criteria, the most significant design competitions across the globe have adopted them for their own judging processes. iF, as the organizer of one of the oldest and most prestigious awards competitions in the field of design, is committed to regularly reviewing these criteria and adjusting them as required to new developments and changing needs. Here's one example: 20 years ago, branding was irrelevant as an evaluation criterion; now it is an essential component of the iF evaluation criteria. iF consistently aims to anticipate change rather than merely reacting to it – which is never more true than for the standards we apply when identifying and honoring outstanding design.

DAN AHN

iF communication design award

Dan Ahn ist Gründer und Geschäftsführer des Intergram Design Studios und Mitbegründer des Saint Paul Arts and Academics. Ahn gewann bereits eine Vielzahl Designpreise, darunter auch mehrere iF awards. Er hält Vorträge zum Thema Design und hat mehrere Bücher veröffentlicht. Ahn war Gastprofessor an der Korea National University of Arts und studierte am Art Center College of Design und am Helsinki School of Economics.

Dan Ahn is the founder and CEO of Intergram Design Studio and co-founder of the Saint Paul Arts and Academics program. He has received many design prizes, including several iF design awards. Dan has authored several books and he is a frequent speaker on design-related issues. He was a visiting professor at the Korea National University of Arts. Dan graduated from the Art Center College of Design (USA) and from the Helsinki School of Economics.

SVEN BAACKE

iF product design award

Sven Baacke (*1974) leitet seit 2011 die Designabteilung Gaggenau Hausgeräte GmbH in München. Dabei verfolgt er einen ganzheitlichen Gestaltungsansatz aus dem Wesen der Marke heraus. Das Designteam ist sowohl in die Produktgestaltung als auch in die Konzeption der gesamten Markenerlebniskette, von Accessoires über Messeauftritte bis hin zu Flagship-Showrooms, eingebunden. Baacke studierte Produktgestaltung an der Staatlichen Akademie der Bildenden Künste in Stuttgart.

Sven Baacke has been head of the design department at Gaggenau Home Appliances (Munich, Germany) since 2011. He focuses on a holistic design approach based on the essence of the brand. The design team is not only involved in product design but also in the concept development for the whole brand experience, including the design of accessories, trade fair exhibitions and flagship showrooms. Sven graduated in product design from the Stuttgart State Academy of Art and Design.

CHRISTOPH BÖNINGER

iF product design award

Christoph Böninger (*1957) studierte Industrial Design in München und Los Angeles. Als Diplomarbeit entwarf er das erste Notebook der Welt, das heute in der Pinakothek der Moderne in München steht. Nach seinem Studium baute er für Siemens in den USA eine Designabteilung auf und war bis 2006 als Designmanager und Geschäftsführer bei Siemens tätig. Daneben entwarf er zahlreiche Möbel, von denen einige in den ständigen Sammlungen von Designmuseen aufgenommen wurden. 2010 gründete er AUERBERG, ein Autoren-Label für Designer und Architekten.

Christoph Böninger studied industrial design in Munich and Los Angeles. As part of his final thesis, he designed the world´s first notebook. After completing his studies, he established a design department for Siemens in the USA and worked as design manager and general manager at Siemens until 2006. He also designed numerous pieces of furniture, some of which have been accepted into the permanent collections of design museums. In 2010, he founded AUERBERG, an author´s label for designers and architects.

RALPH BREMENKAMP

iF communication design award

Ralph Bremenkamp ist Principal Director of Industrial Design bei frog in München. Bremenkamps Kernkompetenz liegt in den Bereichen Consumer Electronics, Wearable Devices, Haushaltsgeräte, Corporate Architecture und Möbeldesign. Vor seiner Zeit bei frog hat Bremenkamp als Designer für Ross Lovegrove in London gearbeitet, als Art Director bei Pixelpark in Berlin sowie als Creative Director für Scholz & Friends. Auf selbstständiger Basis hat er auch mit Firmen in China zusammengearbeitet und viele Produkte für den Haushaltsbereich entwickelt.

Ralph Bremenkamp is principal director of industrial design at frog Munich/Germany. He specialises in the areas of consumer electronics, wearable devices, household appliances, corporate architecture and furniture design. Before joining frog design, he was a designer at Ross Lovegrove's studio in London, he was art director at Pixelpark Berlin and creative director at Scholz & Friends. As a freelancer, he also worked with companies in China and he has developed many household products.

PAUL BUDDE

iF product design award

Paul Budde (*1951) gründete im Jahr 1978 Budde Industrie Design in Münster. Mit seinem Design-Büro ist er in den Bereichen Maschinenbau, technische Konsumgüter sowie Sonderfahrzeuge aktiv. Für Budde ist es wichtig, dass sich bei seinen Arbeiten Form und Funktion vereinen und das Ergebnis einen Mehrwert für Auftraggeber und Konsument darstellt. Budde Industrie Design wurde für die Qualität dieser Arbeit mit zahlreichen Auszeichnungen gewürdigt, darunter drei iF gold awards in den letzten fünf Jahren.

*In 1978, **Paul Budde** opened the Budde Industrie Design office in Münster (Germany) that specializes in mechanical engineering, technical consumer goods and specialized vehicles. In his work, he places great emphasis on uniting form and function and on creating added value for both clients and consumers. Budde Industrie Design has received many awards for the outstanding quality of its work, including, in the last five years, three iF gold awards.*

CONRAD CAINE

iF communication design award

Conrad Caine (*1984) ist Geschäftsführer, Gründer und Inhaber der CONRAD CAINE GmbH in München, einer Full-Service-Medienagentur mit Niederlassungen in Südamerika und mehr als 110 Mitarbeitern. Caine ist auch Gründer der ECKERT & CAINE Software für Management GmbH & Co. KG – ein Unternehmen, das eine spezialisierte Software für strategisches Management entwickelt und liefert, um Führung in komplexen Organisationen zu ermöglichen.

***Conrad Caine** is managing director, founder and owner of the Munich-based CONRAD CAINE GmbH, a full-service digital media agency with headquarters in Germany, subsidiaries in South America and more than 110 employees. Conrad is also founder of ECKERT & CAINE Software for Management GmbH & Co. KG, a company that develops and delivers specialized software for strategic management to enable leadership in complex organizations.*

SERIFE CELEBI

iF product design award

Serife Celebi ist Colour & Material Design Manager bei Ford of Europe Design Studios in Köln. Celebi studierte Textildesign an der Fachhochschule Design & Medien in Hannover mit einem Schwerpunkt auf Stickarbeit und Stricken. Bei Ford ist sie zuständig für Farben und Material in Automodellen wie Focus, C-Max und Kuga. Dabei entwickelt sie Stoffe, Lackierungen, Teppiche, Leder und Farben, die im Auto verwendet werden.

***Serife Celebi** is color & material design manager at Ford of Europe Design Studios in Cologne (Germany). She graduated in textile design from the Hanover University of Applied Sciences where she focused on embroidery and knitting. At Ford, she is in charge of colors and materials for models such as the Focus, the C-Max and the Kuga. She develops fabrics, paints, carpets, leather and colors for use in vehicles.*

NICOLA CHAMBERLAIN

iF product design award

Nicola Chamberlain ist Design-Strategin und Spezialistin im Bereich Colour, Materials and Trend Forecasting bei Veryday in Bromma (Schweden) – ein multidisziplinäres Beratungsunternehmen für Design. Dabei ist sie unter anderem für Electrolux, Motorola, Pepsi und Bombardier tätig. Chamberlains Arbeit umfasst das Forschungsprojekt Lead User Innovation Lab, das in Zusammenarbeit mit dem Interactive Institute in Schweden und IKEA Forschung realisiert wird. Chamberlain hat im Jahr 2000 ihr Studium in Industriedesign abgeschlossen.

***Nicola Chamberlain** is a design strategist and an expert for colors and materials in the trend forecasting team of Veryday, a multidisciplinary design consultancy located in Bromma, Sweden. Her clients include Electrolux, Motorola, Pepsi and Bombardier. She also works on the Lead User Innovation Lab research project, implemented in cooperation with the Interactive Institute in Sweden and with IKEA Research. She graduated in industrial design (2000).*

SHIKUAN CHEN

iF product design award

Shikuan Chen leitet ein Team von über 200 Designexperten bei Compal Electronics, INC. (Taiwan) und verantwortet die Wirtschafts- und Designstrategie des Unternehmens. Daneben ist er als Professor für Design an der Taipei Shih-Chien University und Asia University tätig. Er ist einer der prominentesten Persönlichkeiten der taiwanesischen und chinesischen Design-Industrien. Von 2004 bis 2009 war Chen als Global Design Account Director bei Philips Design (Taipeh/Hongkong) beschäftigt. Er absolvierte 1995 sein Studium Master of Fine Art in Design an der Cranbrook Academy of Art in den USA.

Shikuan Chen is one of the most prominent personalities in Taiwanese and Chinese design. He heads a team of more than 200 design experts at Compal Electronics INC. (Taiwan), where he is responsible for the company's business and design strategies. He is also professor of design at the Shih Chien University (Taipei) and at the Asia University. From 2004 to 2009, he was global design account director at Philips Design (Taipei and Hong Kong). In 1995, he received his MA degree in Fine Art in Design from the Cranbrook Academy of Art (USA).

MARK CHURCHMAN

iF packaging design award

Mark Churchman ist seit 2013 Brand Creative Director beim Küchengerätehersteller Electrolux in Stockholm (Schweden). Dort lenkt er das Kommunikationsdesign der Electrolux Markenfamilie. Zuvor war er bei Philips als Global Creative Director für die Designrichtung der Philips Marken zuständig. Er studierte Grafikdesign und Visual Communications an der Cumbria Institute of the Arts in Carlisle (England).

*Since 2013, **Mark Churchman** has been brand creative director at Electrolux, the Stockholm-based kitchen appliance manufacturer. Mark is responsible for the communication design of the Electrolux brand family. Before taking on the post at Electrolux, he was global creative director at Philips. Mark graduated in graphic design and visual communications from the Cumbria Institute of the Arts in Carlisle (UK).*

PAUL COHEN

iF product design award

Paul Cohen (*1962) ist Mitbegründer von Cube Industrial Design in Sydney (Australien) und Geschäftsführer von Cube Design China – eine der führenden Design Consulting Agenturen im asiatischen-pazifischen Raum. Cohen ist bekannt für sein Design im Bereich Haushaltsgeräte und hat bereits für Marken wie Canon, Electrolux und Johnson & Johnson gearbeitet. Seine Produkte werden in mehr als 50 Ländern verkauft.

Paul Cohen is co-founder of Cube Industrial Design based in Sydney, Australia, and CEO of Cube Design China, one of the leading design consulting agencies in the Asian-Pacific region. Paul is known for his designs in the area of kitchen appliances; he has worked for brands such as Canon, Electrolux and Johnson & Johnson. His products are sold in more than 50 countries.

CHRISTIAN DWORAK

iF communication design award

Christian Dworak ist Creative Director bei Mutabor Design, einer international agierenden Designagentur aus Hamburg, die sich auf Brand Identity spezialisiert hat. Der Schwerpunkt von Christian Dworak liegt in den Bereichen Automotive, Messen und Events, innerhalb derer er mit einem interdisziplinären Team von Architekten, Produktdesignern und Kommunikationsdesignern weltweit Projekte entwickelt und umsetzt. Dworak studierte Kommunikationsdesign an der Fachhochschule Düsseldorf und der Muthesius Hochschule in Kiel.

Christian Dworak is creative director at Mutabor Design, an internationally operating design agency based in Hamburg/Germany and specializing in brand identity. Christian focuses on the areas of automotive design, exhibition and event design. Together with an interdisciplinary team of architects, product designers and communication designers, he develops and implements projects in these areas all over the world. Christian studied communication design at the Düsseldorf University of Applied Sciences and at the Muthesius Academy of Fine Arts and Design in Kiel.

RASMUS FALKENBERG

iF product design award

Rasmus Falkenberg (*1971) ist Mitbegründer des Designstudios design-people in Aarhus (Dänemark). Das 2005 gegründete Designstudio konzentriert sich auf „womenomics" Design, das Frauen als Kunden und Nutzer anspricht. Falkenberg absolvierte 2001 sein Masterstudium in Industriedesign an der Aarhus School of Architecture.

Rasmus Falkenberg is co-founder of the design-people studio based in Aarhus, Denmark. The studio was founded in 2005 and specializes in 'womenomics', i.e. in developing concepts that specifically take into consideration female users and clients. In 2001, Rasmus received his MA in industrial design from the Aarhus School of Architecture.

KHODI FEIZ

iF product design award

Khodi Feiz (*1963) ist Mitbegründer des Feiz Design Studio in Amsterdam (Niederlande), das auf Produkt-, Möbel-, Grafik- und Strategiedesign spezialisiert ist. Der Industriedesigner, der im Iran geboren wurde, in den USA aufwuchs und mittlerweile in den Niederlanden lebt, hatte zuvor bei Philips Design im Bereich Advanced Design Group als Designmanager und kreativer Direktor gearbeitet

Khodi Feiz is an Iran-born industrial designer who was raised in the USA and now lives in Amsterdam (The Netherlands) where he co-founded the Feiz Design Studio. The studio specializes in product design, furniture design, graphic design and strategic design. Before opening his own studio, Khodi was design manager and creative director at Philip's Advanced Design Group.

JENNY FLEISCHER

iF packaging design award

Jenny Fleischer ist seit 2010 Global Head of Design Management beim Hautpflegeunternehmen Beiersdorf in Hamburg mit dem Ziel, Design als Erfolgsfaktor der wirtschaftlichen Strategie zu fördern. Außerdem verantwortet sie den Entwicklungsprozess und die Markteinführung der neuen Designsprache der Marke Nivea und fördert marken-gesteuerte Innovation.

Since 2010, ***Jenny Fleischer*** *has been global head of design management at the German skin care products manufacturer Beiersdorf in Hamburg (Germany). Her goal is to promote design as a success factor in business strategy. Jenny is also responsible for the development and launch of the new design for the Nivea brand and for brand-driven innovation.*

PAUL FLOWERS

iF product design award

Paul Flowers (*1972) schloss sein Studium an der Universität von Northumbria, Newcastle, mit Auszeichnung ab. Seine facettenreiche internationale Karriere führte ihn von angesehenen Londoner Design-Studios bis hin zu einigen der größten Unternehmen der Welt. Er hat u. a. Projekte für FM Design und IBM in Großbritannien, Electrolux in Italien und Philips in den Niederlanden entworfen. Seit 2005 ist er beim Sanitärarmaturen-Hersteller GROHE Global Senior Vice President of Design.

*****Paul Flowers*** *graduated with an honors degree from the University of Northumbria, Newcastle (UK). His multi-facetted international career has included posts both at renowned London-based design studios and at some of the largest international companies. Paul has carried out design projects for, among others, FM Design and IBM (UK), Electrolux (Italy) and Philips (The Netherlands). Since 2005, he has been global senior vice president of design at GROHE (Germany), a leading provider of premium bathroom fittings.*

PROF. FRITZ FRENKLER

Chairman iF design awards

Prof. Fritz Frenkler (*1954) studierte Industrial Design an der HBK Braunschweig. Er arbeitete als Geschäftsführer von frogdesign Asien, leitete die wiege Wilkhahn Entwicklungsgesellschaft und war Design-Chef der Deutschen Bahn AG. Im Jahr 2000 gründete er mit Anette Ponholzer die f/p design deutschland gmbh und 2003 f/p design japan inc. Frenkler ist Regional Advisor des ICSID, seit vielen Jahren Juryvorsitzender des iF product design award und Gründungsmitglied des universal design e.V., Hannover. 2006 wurde er als Ordinarius auf den neuen Lehrstuhl für Industrial Design an der Technischen Universität München (TUM) berufen. Seit 2013 ist er Mitglied der Akademie der Künste in Berlin.

Prof. Fritz Frenkler studied industrial design at the Hochschule für Bildende Künste in Braunschweig (Germany). He was general manager at frogdesign Asia, general manager of wiege Wilkhahn Entwicklungsgesellschaft and head of design at Deutsche Bahn AG. In 2000, together with Anette Ponholzer, he founded f/p design deutschland gmbh followed, in 2003, by f/p design japan inc. Fritz is a regional advisor of the ICSID; he has been jury chairman for the iF product design award for many years and he is a founding member of universal design e.V., Hannover. In 2006, he was appointed tenured professor of the new faculty of industrial design at the Technische Universität München (TUM).

PATRICK FREY

iF product design award

Patrick Frey wurde in Seoul/Südkorea geboren. 2007 gründete er das Designbüro Patrick Frey Industrial Design, das für Firmen wie Nils Holger Moormann, Bree, Authentics, Richard Lampert, Elmar Flötotto, FreiFrau und andere arbeitet. Schon im ersten Jahr der Selbstständigkeit erhielt Patrick Frey den höchstdotiertesten Preis für Nachwuchsdesigner, den Lucky Strike Junior Designer Award. Zahlreiche seiner Produkte erhielten internationale Preise, wie den iF product design award, Good Design Award Chicago Athenaeum, Designpreis Baden – Württemberg, Interior Innovation Award. Patrick Frey (v. Prof.) lehrt an der Hochschule Hannover Produktdesign.

Patrick Frey was born in Seoul, South Korea. In 2007, he founded the Patrick Frey Industrial Design studio, working with clients such as Nils Holger Moormann, Bree, Authentics, Richard Lampert, Elmar Flötotto, FreiFrau and others. In the first year of being an independent designer, Patrick Frey already received the most valuable Lucky Strike Junior Design Award. Many of his products received international prizes such as the iF product design award, the Good Design Award Chicago Athenaeum, the Designpreis Baden- Württemberg and the Interior Innovation Award. Patrick Frey is professor of product design at the Hannover University.

NEIL GRIDLEY

iF product design award

Neil Gridley ist als Design Management Consultant für Unternehmen wie Philips, Electrolux, Unilever und dem Design Council tätig. Er berät auch Viadynamics, ein in London ansässiges Beratungsunternehmen für Blue Chip-Kunden sowie aufstrebende Start-ups. Als Design Associate of the Design Council ist er verantwortlich für mehrere Programme, um kleinen und mittleren Unternehmen, High Tech Start-Ups und Universitäten dabei zu helfen, ihr Design zu stärken. Er studierte Industriedesign und schrieb mit Yasushi Kusume (Vice President of Design bei Electrolux) das Buch „Brand Romance".

Neil Gridley is a Design Management Consultant working with clients such as Philips, Electrolux, Unilever and the Design Council. He also consults for Viadynamics, a London based Innovation Consultancy for International Blue Chip clients as well as incubating start-ups. As a Design Associate of the Design Council he is responsible for several programs devised to help SMEs, high tech start-ups and universities with strengthening their design. Neil graduated in industrial design and, together with Yasushi Kusume (vice president of design at Electrolux), he co-authored 'Brand Romance', a book on building loved brands.

LARS LYSE HANSEN

iF product design award

Lars Lyse Hansen ist Geschäftsführer der Bolia.com International in Aarhus (Dänemark) – ein Möbelhersteller, der mit mehr als 70 Designern zusammenarbeitet. Gemeinsam mit ihnen entwirft und entwickelt die Möbelkette ein breites Sortiment an Möbeln und Einrichtungszubehör. Bevor Hansen 2005 Bolia.com gründete, war er zehn Jahre im Management des schwedischen Einrichtungshauses Ikea tätig. Er hält Vorträge im In- und Ausland. Zu seinen größten Leidenschaften gehören das Bergsteigen und die Musikproduktion.

Lars Lyse Hansen is CEO of Bolia.com International based in Aarhus (Denmark), a furniture manufacturer working together with more than 70 designers in developing a broad spectrum of furniture and interior fittings. Before founding Bolia.com in 2005, Lars was a senior manager with Swedish furniture retailer IKEA. He is a frequent international speaker. Besides the job Lars is a passionate mountaineer, music producer and familianaire.

SAM HECHT

iF product design award

Sam Hecht gründete das Designbüro Industrial Facility (London) mit seiner Partnerin Kim Colin. Gemeinsam haben sie Projekte für Unternehmen wie Mattiazzi, Issey Miyake, Yamaha und Established & Sons entwickelt. Sam Hecht war als Kreativberater für Muji und das US-amerikanische Unternehmen Herman Miller tätig. Das MoMA in New York und das Centre Pompidou in Paris haben Arbeiten von Hecht in ihre Sammlungen aufgenommen, und Industrial Facility wurde vor kurzem vom Forbes Magazine zu einem der zehn progressivsten Industriedesignbüros der Welt ernannt.

Sam Hecht is co-founder of the Industrial Facility design office (London) which he runs together with his partner Kim Colin. The office has developed production projects for companies such as Mattiazzi, Issey Miyake, Yamaha and Established & Sons. As a creative advsior, Sam has worked for Muji and Herman Miller Co. of America. His work is included in the permanent collections of the MoMA, New York, and of the Centre Pompidou, Paris. Forbes Magazine recently listed his studio among the ten most progressive industrial design offices in the world.

FRED HELD

iF product design award

Fred Held (*1966) studierte und diplomierte an der Muthesius Hochschule in Kiel im Bereich Industrie Design. Nach einigen Jahren als freier Mitarbeiter bei Windi Winderlich Design in Hamburg unterstützte er 1997 die KIDP (Korean Institute of Design Promotion) in Süd-Korea bei der Projektarbeit. Er gründete im selben Jahr Held+Team in Hamburg und spezialisierte sich auf Industriedesign für Medizintechnik. Im Jahre 2011 gründete er Held+Team Design Partnergesellschaft gemeinsam mit Thomas Märzke und ist seitdem weltweit für Medizintechnikunternehmen tätig.

Fred Held graduated from the Muthesius Academy in Kiel (Germany) with a degree in industrial design. In 1997, after some years working as a freelancer for Windi Winderlich Design in Hamburg, he provided project support to the KIDP (Korean Institute of Design Promotion) in South Korea and he founded Held+Team, a Hamburg-based design office specializing in industrial design for medical technology. In 2011, together with Thomas Märzke, he founded Held+Team Design Partnergesellschaft. Fred works for medical technology companies all over the world.

TOM HIRT

iF communication design award

Tom Hirt studierte Produktgestaltung (Schwerpunkt digitale Medien) an der HTW Dresden. Während des Studiums arbeitete er parallel bei Designagenturen, Architekturbüros und Firmen. 1998 übernahm er als Projektleiter Internet/Intranet den Bereich der digitalen Medien bei der ERCO Leuchten GmbH in Lüdenscheid. Zwei Jahre später die Gesamtleitung der Digitalen Kommunikation und zusätzlich (2004-2005) die User Interface Design Entwicklung im Bereich Forschung und Entwicklung. Seit 2002 lehrt er an der FH Düsseldorf im Fachgebiet Interaktive Systeme/Hypermedia das Thema Interface- und Interaktionsgestaltung. Daneben führt er Projektarbeiten, leitet Workshops und hält Vorträge.

Tom Hirt studied product design (focusing on digital media) at the HTW Dresden/Germany. Alongside his studies, he also worked at design agencies, at architectural practices and at other companies. In 1998, he joined the digital media department of ERCO Leuchten GmbH (Lüdenscheid, Germany) as the project lead for Internet/Intranet. Two years later, he took over the post as director of digital communication and, from 2004 o 2005, he also headed ERCO's user interface design team in the area of R&D. Since 2002, he has been teaching interface and interaction design at the Düsseldorf University of Applied Sciences' department of Interactive Systems/Hypermedia. He also supervises project work, holds workshops and gives talks.

DANIELLE HUTHART

iF communication design award

Danielle Huthart ist Inhaberin des Designstudios Whitespace in Hong Kong, das Kreativdirektion und Services im Bereich Branding bietet. Das Studio entwickelt Identitäten für Marken in der Retail-und-Fashion-Branche sowie in der Kunst- und Kulturindustrie. Außerdem ist sie als Art Director bei der Zeitschrift ArtAsiaPacific tätig.

Danielle Huthart is the owner of the Whitespace design studio in Hong Kong that provides services in the areas of art direction and branding. The studio develops CIs for brands in the retail and fashion sectors and in the art and culture industries. Danielle is also the art director of the ArtAsiaPacific magazine.

RONALD IHRIG

iF product design award

Ronald Ihrig studierte Industriedesign in Darmstadt und sammelte 1982 in den Studios der Adam Opel AG erste praktische Erfahrungen im Automobildesign. 1984 gründete er sein eigenes Designstudio in der Nähe von Frankfurt. Ihrig kommentiert Trends der internationalen Automobilausstellungen und lehrt seit 2010 Automobilgeschichte an der Hochschule Reutlingen. Daneben organisiert er die Automotive Designers'Night in Frankfurt und Paris und ist Jurymitglied diverser internationaler Designwettbewerbe.

Ronald Ihrig studied industrial design in Darmstadt, Germany. In 1982, he gained his first practical experience in automotive design in the studios of the Adam Opel AG. In 1984, he founded his own design studio based near Frankfurt. Ronald is a frequent commentator on the latest trends from the major international automotive exhibitions and, since 2010, he has been a lecturer in automotive history at the University of Reutlingen. He also organizes the Automotive Designers'Night in Frankfurt and Paris and he is a juror in many international design competitions.

PETER IPPOLITO

iF communication design award

Peter Ippolito (*1967) studierte Architektur in Stuttgart und Chicago. Während dieser Zeit war er als Assistent von Prof. Ben Nicholson (Chicago) tätig und sammelte praktische Erfahrungen im Studio Daniel Libeskind (Berlin). 1999 war er Gründungsmitglied von zipherspaceworks. Aus diesem Büro ging 2002 die Ippolito Fleitz Group / identity architects hervor, die er seitdem gemeinsam mit Gunter Fleitz führt.

Peter Ippolito studied architecture in Stuttgart and Chicago. During his student years, he worked as an assistant to Prof. Ben Nicholson (Chicago) and he gained practical experience at Studio Daniel Libeskind (Berlin). In 1999, he was among the founding members of zipherspaceworks. In 2002, zipherspaceworks became Ippolito Fleitz Group / identity architects, a company Peter runs together with Gunter Fleitz.

YOSHIFUMI ISHIKAWA

iF product design award

Yoshifumi Ishikawa (*1961) begann 1984 nach seinem Studium an der Tamagawa Universität seine Karriere bei Canon Inc. in Tokio (Japan) und gestaltete Kameras. 1998 wechselte er zu Canon USA als Senior Business Planner im Internet-Geschäft. Nach seiner Rückkehr nach Japan arbeitete Ishikawa als Advanced User Interface Designmanager und als Leiter der Interface Design Abteilung. Seit 2012 leitet er als Senior General Manager das Design Center. Neben dem klassischen Produktdesign beschäftigt er sich derzeit mit der Verwirklichung visionärer Zukunftsthemen.

*After graduating from the Tamagawa University in 1984, **Yoshifumi Ishikawa** started his career at Canon Inc. (Tokyo, Japan) where he designed cameras. In 1998, he went to Canon USA to take on the post of senior business planner for online business. After his return to Japan, Yoshifumi became advanced user interface design manager and head of interface design. Since 2012, he has been heading Canon's design center as senior general manager. In addition to traditional product design, one of his recent interests is the implementation of visionary, future-oriented concepts.*

RONALD KAPAZ

iF communication design award

Ronald Kapaz (*1956) ist Mitbegründer der Designagentur Oz Strategy + Design. Kapaz ist als Seniorberater unter anderem für den Bereich Corporate Identity tätig und hat für Kunden wie Coca-Cola, Credit Suisse, Unilever, McDonald's gearbeitet. Er veröffentlichte diverse Publikationen über Design und ist als nationaler und internationaler Vortragsredner und Juror tätig.

***Ronald Kapaz** is co-founder of the Oz Strategy + Design agency. As a senior consultant, his work includes corporate identity projects and he has worked with clients such as Coca-Cola, Credit Suisse, Unilever and McDonald's. He has authored several publications on design and he is a frequent speaker and juror at home and abroad.*

SULTAN KAYGIN SEL

iF product design award

Sultan Kaygin Sel ist Industriedesignerin mit langjähriger Erfahrung in Industrie und Forschung. Sie erhielt ihren Doktortitel von der METU (Middle East Technical University) in Ankara (Türkei), wo sie von 2001 bis 2003 als wissenschaftliche Mitarbeiterin tätig war. Seit 2003 arbeitet sie bei Vestel Electronics (Türkei). Für ihre Arbeit wurde sie bereits mit mehreren Designpreisen ausgezeichnet und sie war ebenfalls als Jurymitglied für verschiedene Preise tätig. Sie verfügt über langjährige Erfahrung im Industriedesign und im Projektmanagement für LED TV-Geräte, All-in-One PCs, Fernbedienungen, Satellitenempfängern, Mobiltelefonen sowie LED Beleuchtung. Sie war auch schon im Bereich User Interface (R&D research studies) aktiv.

Sultan Kaygin Sel is an industrial designer who has combined her professional design experience with academic studies. She has received her PHD degree from METU (Middle East Technical University) in Ankara (Turkey). She worked as a research assistant at METU between 2001 and 2003. Since 2003, she has been working at Vestel Electronics (Turkey). During her award winning design career she has acted as a jury member in different design competitions. She has extensive experience in industrial design and project management of LED TVs, All-in-One PCs, remote controls, satellite receivers, mobile phones and LED Lighting. She is also experienced in user interface R&D research studies.

HARRI KOSKINEN

iF product design award

Harri Koskinen ist seit 2012 Designchef des finnischen Glas- und Geschirrherstellers iittala. Nach seinem Studium an der Alvar Aalto Universität in Helsinki (Finnland) gründete er im Jahr 2000 die Designagentur Friends of Industry Ltd. 2009 brachte Koskinen zudem seine erste Möbelkollektion auf den Markt: Harri Koskinen Works. Im selben Jahr wurde er Partner bei Eat&Joy Farmer's Market, einer finnischen Nahrungsmittelkette für organische und regionale Produkte. Koskinens Arbeiten, werden auf Ausstellungen in aller Welt gezeigt, unter anderem die Leuchte Block im MoMa.

*Since 2012 **Harri Koskinen** has been head of design at Finnish tableware manufacturer iittala. After graduating from the Alvar Aalto University in Helsinki (Finland), in 2000, he founded the design agency Friends of Industry Ltd. In 2009, he launched his first furniture collection: Harri Koskinen Works. Also in 2009, he became a partner in Eat&Joy Farmer's Market, a Finnish chain selling organic and regional products. Koskinen's work is presented in exhibitions all over the world: his Block lamp, for example, is included in the MoMa collection.*

FREEMAN LAU

iF packaging design award

Freeman Lau (*1958) ist Partner der Kan & Lau Design Consultants in Hongkong. Freemans Design ist umfangreich: es umfasst Buch-, Graphik- und Kartengestaltung. Freeman setzt sich für Design ein, indem er an öffentlichen Design-Veranstaltungen teilnimmt. Er war beispielsweise Vorsitzender der Hong Kong Designers Association und unterstützt damit das Potenzial der Designindustrie in Hong Kong. Er ist Absolvent der Hong Kong Polytechnic.

Freeman Lau is a partner in the Hong Kong-based Kan & Lau Design Consultants firm. His work ranges from book design and graphic design to card design. Freeman supports the cause of design by frequently participating in public design events. He was, for example, chairman of the Hong Kong Designers Association, one of the activities by which he supports the potential of Hong Kong's design industry. He graduated from the Hong Kong Polytechnic University.

MONIKA LEPEL

iF communication design award

Seit 1993 führt **Monika Lepel** gemeinsam mit ihrem Mann das Büro LEPEL & LEPEL Architektur, Innenarchitektur in Köln. An der Peter Behrens School of Architecture, Düsseldorf, unterrichtete sie vier Jahre lang Gestaltungsgrundlagen. Zuvor war sie nach ihrer Ausbildung in Düsseldorf und Salzburg als leitende Innenarchitektin bei KSP Köln tätig. „Beziehungen bauen" ist ihr zentrales Thema. Als Innenarchitektin entwirft Monika Lepel mit ihrem Team unverwechselbare Räume für Unternehmen. Die Arbeiten wurden mit zahlreichen renommierten Preisen ausgezeichnet.

*Since 1993, **Monika Lepel** has been heading the Cologne-based LEPEL & LEPEL Architektur, Innenarchitektur office, which she runs together with her husband. For four years, she taught at the Peter Behrens School of Architecture (fundamentals of design) in Düsseldorf. Before that, she was a senior interior designer at KSP Köln, based at the company's offices in Düsseldorf and Salzburg. Her central theme is 'building relationships'. As an interior designer, Monika Lepel and her team create unique interiors for companies. Her work has received many renowned prizes.*

STEVE LEUNG

iF communication design award

Steve Leung ist Architekt, Innenarchitekt und Produktdesigner. Im Jahr 1987 eröffnete er sein Beratungsbüro für Architektur und Stadtplanung, das er 1997 in „Steve Leung Designers Ltd." umstrukturierte. Darüber hinaus gründete Leung die auf Gastronomie Management spezialisierte Lifestyle Marke „1957 & Co.". 2013 wurde er von dem international bekannten Designlabel „yoo" als Creative Director für „Steve Leung & yoo" akquiriert. Seine Arbeiten wurden bereits mit zahlreichen Preisen ausgezeichnet.

Steve Leung is an architect, interior designer and product designer. He set up his own architectural and urban planning consultancy in 1987 and restructured the company into Steve Leung Designers Ltd. in 1997. Apart from that, Steve founded the 1957 & Co. lifestyle brand that focuses on hospitality management in 2007. In 2013, the internationally renowned design company yoo offered him the post of Creative Director of 'Steve Leung & yoo'. Steve has received many awards and has been a jury member for several renowned design competitions.

TOMMY LI

iF communication design award

Tommy Li ist weltweit als Branding Designer tätig. Seine Agentur Tommy Li Design Workshop ist eine der führenden Agenturen für Branding in China. Li arbeitet für eine Vielzahl chinesischer und internationaler Kunden wie Swarovski, MTR Corporation, Honeymoon Dessert und Dairy Farm Group. Er studierte an der School of Design in der Hong Kong Polytechnic University und erhielt schon über 580 Auszeichnungen.

Tommy Li is a branding designer who works on an international level. The Tommy Li Design Workshop is one of the leading branding agencies in China. Its clients include many Chinese and international companies such as Swarovski, MTR Corporation, Honeymoon Dessert and Dairy Farm Group. He graduated from the Hong Kong Polytechnic University's School of Design and he has received more than 580 awards.

LIDAN LIU

iF product design award

Lidan Liu (*1975) ist Geschäftsführerin von designaffairs China – das asiatische Headquarter der inhabergeführten, strategischen Design Consulting Agentur mit Hauptsitz in München. Liu studierte Industriedesign an der Universität Essen. Danach arbeitete sie unter anderem für das MEDION design center Germany, XLPLUS Design und die Folkwang Universität.

Lidan Liu is CEO of designaffairs China, the Asian branch of the owner-led strategic design consulting agency headquartered in Munich, Germany. Lidan graduated from the University of Essen with a degree in industrial design. She worked for, among others, the MEDION design center Germany, XLPLUS Design and the Folkwang University.

SEBASTIAN MAIER

iF product design award

Sebastian Maier (*1970) ist seit 2007 geschäftsführender Gesellschafter der Corpus-C Design Agentur in Fürth mit dem Schwerpunkt Labor-, Pharma- und Medizintechnik. Bevor er 2002 in die Medizintechnik bei Siemens Medical in Erlangen einstieg, studierte er Maschinenbau und Produktdesign in Braunschweig. Zu den heutigen Auftraggebern zählen renommierte Unternehmen wie Siemens Medical, Dräger Medical, KUKA Laboratories und die Sartorius AG.

Since 2007, Sebastian Maier has been CEO of the Corpus-C Design agency based in Fürth (Germany) and specializing in technologies for the pharmaceutical, laboratory and medical sectors. Before joining Siemens Medical (Erlangen, Germany) in 2002, he studied mechanical engineering and product design in Braunschweig. His clients include renowned companies such as Siemens Medical, Dräger Medical, KUKA Laboratories and Sartorius AG.

JUNKI MOON

iF product design award

Junki Moon ist seit 1996 Geschäftsführer bei M.I. DESIGN INC. (Südkorea). Seine Agentur hat sich auf Beratung im Bereich Produktdesign und Service-Design spezialisiert. Zum Kundenstamm gehören Unternehmen wie LG Electronics, SK chemicals, Yujin Robot und Soosan. Seit 2012 ist Moon Vizevorsitzender der Korea Federation of Design Associations (kfda). Von 2007 bis 2011 war Moon als Professor am Samsung Art and Design Institute tätig.

Junki Moon has been CEO of M.I. DESIGN INC. (South Korea) since 1996. His agency specializes in consulting services for the product design and service design sectors. His clients include companies such as LG Electronics, SK chemicals Yujin Robot and Soosan. Since 2012, Junki has been chairman of the Korea Federation of Design Associations (kfda). From 2007 to 2011, he was a professor at the Samsung Art and Design Institute.

ACHIM NAGEL

iF product design award

Achim Nagel (*1959) studierte Architektur an der Technischen Universität Hannover. Nach seiner Tätigkeit in dem Architekturbüro Schweger und Partner war er von 1988 bis 1993 Leiter der Zentralen Bauabteilung der Bertelsmann AG in Gütersloh. Von 1993 bis 1999 war er Partner im Architekturbüro Ingenhoven Overdiek + Partner in Düsseldorf, bevor er dort im Jahr 2000 die PRIMUS Immobilien AG gründete, gefolgt vom 2005 gegründeten möser projekt GbR mit Jürgen Möser. 2009 gründete er mit Christian F. Heine die ACP Planungsgesellschaft mbH & Co. KG.

Achim Nagel studied architecture at the University of Hannover. From 1988 to 1993, he was the head of the central construction department at Bertelsmann AG (Gütersloh, Germany). From 1993 to 1999, Achim was a partner of the Ingenhoven Overdiek + Partner architectural practice in Düsseldorf before founding PRIMUS Immobilien AG and PRIMUS developments GmbH in 2000 and 2001 respectively (both based in Düsseldorf). In 2005, together with Jürgen Möser, he founded möser projekt GbR and in 2009, together with Christian F. Heine, he founded the ACP Planungsgesellschaft mbH & Co. KG.

SEYHAN ÖZDEMIR

iF product design award

Seyhan Özdemir (*1975) gründete 2003 gemeinsam mit ihrem Partner Sefer Cağlar das Designstudio Autoban in Istanbul (Türkei), das sich auf Architektur- und Interieurprojekte sowie Produktdesign spezialisiert hat. Seit 2007 arbeitet Autoban mit dem englischen Designstudio De La Espada zusammen: Autoban ist stärker am kreativen Entwicklungsprozess beteiligt, während De La Espada die Entwürfe durch Handwerkskunst realisiert. Autobans Interieur- sowie Produktentwürfe haben bis heute mehr als 15 internationale Auszeichnungen erhalten. Özdemir studierte Architektur an der Fine Arts Faculty der Mimar Sinan University.

*In 2003, **Seyhan Özdemir** founded the Autoban design studio (Istanbul, Turkey) together with her partner Sefer Cağlar. The studio specializes in architecture and interior design projects and in product design. Since 2007, Autoban has been cooperating with the UK design studio De La Espada: while Autoban focuses on the creative development process, De La Espada specializes in craftsmanship and the implementation of the designs. Autoban has received more than 15 international awards for their interior design projects and product designs. Seyhan graduated in architecture from the Mimar Sinan University's fine arts faculty.*

OSCAR PEÑA

iF product design award

Oscar Peña wurde in Kolumbien geboren und besuchte die Xavier University in Bogota, Kolumbien, wo er 1982 sein Diplom als Industrial Designer erhielt. Er ist bei Philips Design Lighting als Global Creative Director, Product and User Experience Design, tätig. Er überwacht Designprojekte und strategische Designinitiativen, insbesondere für Produkte im Bereich Beleuchtung. Unabhängig davon, ob die Gestaltung auf einer konzeptionellen Ebene oder für den Massenmarkt erfolgt, ist Peñas Erforschen „der Essenz des ernsten Spiels" ebenso ein wiederkehrendes Thema in seiner Arbeit wie sein Fokus auf den menschlichen Inhalt zeitgenössischen Designs.

***Oscar Peña** was born in Colombia and graduated from the Xavier University in Bogota, Colombia, with a degree in industrial design (1982). He is the global creative director, product and user experience design, at Philips Design Lighting. Oscar oversees both everyday design projects and strategic design initiatives, particularly for lighting products. Whether designing on a conceptual level or for the mass market, Oscar's investigation of 'the essence of serious play' and his focus on the human aspect of contemporary design are recurring themes in his work.*

FLORIAN PETERSEN

iF communication design award

Florian Petersen ist Gründer der Agentur cgi STUDIO GmbH in Berlin, in der er als Geschäftsführer und Creative Director tätig ist. Der Schwerpunkt der Firma liegt auf Produktvisualisierung von digitalen Daten. Zum Kundenstamm gehören Konzerne wie Bosch, Siemens, Gaggenau und Neff. Petersen sammelte erste Berufserfahrungen beim Film, dort hat er fundamentale Kenntnisse aus den Bereichen Licht und Kamera sowie digitaler Bildgestaltung in 3D erlernt. Danach war er fünf Jahre im Design bei Volkswagen tätig.

***Florian Petersen** is the founder, CEO and creative director of cgi STUDIO GmbH in Berlin (Germany). The agency focuses on the product visualization of digital data. Clients include companies such as Bosch, Siemens, Gaggenau and Neff. Florian started his career in the film industry where he gained profound skills in the areas of lighting, cinematography and digital 3D visualization. After leaving the film industry, he worked with Volkswagen for five years.*

PROF. VOLKER POOK

iF communication design award

Volker Pook (*1969) ist seit 2008 Professor an der BTK/Hochschule für Gestaltung in Berlin. Nach seinem Designstudium an der Gesamthochschule Universität in Essen (heute Folkwang Universität der Künste), arbeitete er zunächst als Designer bei Büro Hamburg/Hamburg. Es folgten Engagements bei MetaDesign und den Peter Schmidt Studios. Von 2001 bis 2004 wirkte er in New York bei Agenturen wie Enterprise IG (heute The Brand Union) und Wolff Olins an Corporate Identity-Projekten für Kunden wie Ford Motor Company, General Electric und Staples als Senior Designer mit.

*Since 2008, **Volker Pook** has been a professor at the BTK University of Applied Sciences (Berlin, Germany). After graduating in design from the Essen Gesamthochschule Universität (today: Folkwang University) he worked as a designer at the Hamburg/Hamburg office, followed by posts at MetaDesign and at Peter Schmidt Studios. From 2001 to 2004, he was a senior designer at New York based agencies such as Enterprise IG (today: The Brand Union) and Wolff Olins, where he cooperated in CI projects for clients including Ford Motor Company, General Electric and Staples.*

ANDRÉ POULHEIM

iF product design award

André Poulheim (*1974) gründete 2001 gemeinsam mit Thorsten Frackenpohl das Designstudio Frackenpohl Poulheim in Köln. Bei der Entwicklung von Produkt- und Dienstleistungslösungen ist es ihm wichtig, die Nutzer-Perspektive nie aus den Augen zu verlieren. Im Rahmen eines Lehrauftrags an der FH Köln beschäftigt er sich mit dem Potenzial von Produkt-Service-Systemen. Er studierte an der Köln International School of Design.

André Poulheim is co-founder of the Frackenpohl Poulheim design studio located in Cologne, Germany (founded in 2001, together with Thorsten Frackenpohl). When developing both product and service solutions, he always bears the user perspective in mind. In the context of a research project at the Cologne University of Applied Sciences, he looks into the potential of product-service systems. He graduated from the Köln International School of Design.

ROBERT SACHON

iF product design award

Robert Sachon studierte zunächst an der Freien Kunstschule in Mannheim, anschließend an der FH Darmstadt, wo er 1998 seinen Abschluss als Diplom Industriedesigner machte. Seit 1999 arbeitete er als Designer bei der Siemens Elektrogeräte GmbH. Ab 2001 leitete Robert Sachon die zentrale Designabteilung für die zur Bosch und Siemens Hausgeräte GmbH gehörenden Regionalmarken Balay, Lynx (Spanien), Pitsos (Griechenland), Profilo (Türkei), Continental (Brasilien) und Coldex (Peru). 2005 wurde er Chefdesigner und ist jetzt Global Design Director der Robert Bosch Hausgeräte GmbH und gewann seither zahlreiche Designpreise.

Robert Sachon first studied at the Independent School of Arts in Mannheim/Germany and then at Darmstadt University of Applied Sciences, where he graduated as an industrial designer in 1998. He worked as a designer at Siemens Elektrogeräte GmbH starting in 1999. Beginning in 2001, Robert Sachon headed the central design department for the Bosch and Siemens Hausgeräte GmbH regional brands Balay, Lynx (Spain), Pitsos (Greece), Profilo (Turkey), Continental (Brazil) and Coldex (Peru). He became Head Designer and is now Global Design Director of Robert Bosch Hausgeräte GmbH in 2005 and has won numerous design awards since then.

STEFAN SCHOLTEN

iF product design award

Stefan Scholten (*1972) gründete im Jahr 2000 gemeinsam mit Carole Baijings in Amsterdam (Niederlande) das Scholten & Baijings Studio for Design. Mittlerweile ist das Designerduo fester Bestandteil des niederländischen Designs. Ihre Produkte für den Innenbereich überzeugen durch viele neue Ansätze. Scholten ist Absolvent der Design Academy in Eindhoven.

*In 2000, together with Carole Baijings, **Stefan Scholten** founded the Scholten & Baijings Studio for Design in Amsterdam (The Netherlands), which has become an established name in Dutch design. The Scholten & Baijing interior design products stand out by their innovative qualities. Stefan graduated in design from the Design Academy Eindhoven.*

BRIAN STEPHENS

iF product design award

Brian Stephens (*1955) gründete 1984 die Product Design Consultancy Design Partners in Dublin (Irland). Nach seinem Studium an der Birmingham Polytechnic arbeitete er bei Ogle Design in Großbritannien. Von dort wechselte er zu GK Design in Tokio und konzentrierte sich dort auf Projekte im Bereich Verkehr und Investitionsgüter. Die Zeit in Japan beeinflusst seine kreative Arbeit und den Arbeitsstil seines Designstudios.

*In 1984, **Brian Stephens** founded the Design Partners consultancy firm for product design, based in Dublin (Ireland). After graduating from Birmingham Polytechnic, he worked at Ogle Design (UK). He then moved to Tokyo to work with GK Design where he focused on transportation and capital goods projects. Both his own and the studio's creative work are influenced by his time in Japan.*

MARTIN TOPEL

iF product design award

Martin Topel lehrt seit 1999 als Professor an der Universität Wuppertal im Studiengang Industrial Design. Sein Lehrstuhl beschäftigt sich mit der Produktentwicklung von Investitionsgütern und Produktsystemen. Martin Topel entwickelte eine Reihe von Produkten für Unternehmen wie Bosch, Festool und mehrere internationale Airlines. Im Jahr 2010 gründete er zusammen mit einem Partner die Agentur Squareone GmbH in Düsseldorf. Dort beschäftigt er sich neben der klassischen Produktentwicklung für Mittelständler und Konzerne zunehmend mit der Beratung innerhalb der strategischen Innovationsberatung.

*Since 1999, **Martin Topel** has been professor of industrial design at the University of Wuppertal (Germany) where he focuses on the development of investment goods and product systems. He has developed products for companies such as Bosch and Festool and for several international airlines. In 2010, he co-founded the Squareone GmbH agency in Düsseldorf. In addition to traditional product development for both SMEs and large companies, the agency provides consulting services in the area of strategic innovation.*

JAMES TURNER

iF communication design award

James Turner ist seit 2008 Art Director für Kommunikationsdesign beim Sony Design Center Europa. Sein Team und er arbeiten weltweit an Projekten, von Markenentwicklung über Verpackungsdesign bis hin zu Digital Entertainment und Advertising. Turner studierte an der Swansea Metropolitan University in Wales und begann seine Karriere mit dem Design von Katalogen für das Interior Design Unternehmen OKA. 2004 wechselte er zum Modeunternehmen Harvey Nichols, wo er für das Graphikdesign zuständig war.

*Since 2008, **James Turner** has been art director of communication design at the Sony Design Center Europe. Together with his team, he works on international projects ranging from brand development and packaging design to digital entertainment and advertising. James graduated from the Swansea Metropolitan University in Wales (UK) and started his career with designing catalogs for the interior design company OKA. In 2004, he took on a post at the luxury fashion retailer Harvey Nichols where he was responsible for the company's graphic design.*

OSKAR ZIETA

iF product design award

Oskar Zieta (*1975) ist Architekt und Designer. Im Jahr 2008 lancierte Zieta den preisgekrönten Hocker PLOPP – ein Produkt, das durch die sogenannte FIDU-Technologie („Freie InnenDruck Umformung") entsteht. Seit 2011 leitet er das Department Industrial Design an der SOF (school of form) in Poznan (Polen) und bietet dort Workshops zu den Themen Architektur und Design an. Im Jahr 2010 gründete er Zieta Prozessdesign, eine zwischen Architektur, Design und Leichtbau interdisziplinär tätige Firma.

Oskar Zieta is an architect and designer. In 2008, he launched the award-winning PLOPP stool, a product created by using the FIDU (Free Internal Pressure Reshaping) technology. Since 2011, he is the director of the industrial design department at the SOF (school of form) in Poznan, Poland where he offers workshops in architecture and design. In 2010, he founded Zieta Prozessdesign, a company that works in an interdisciplinary way by combining the areas of architecture, design and lightweight construction.

Manufacturer/Client **1**

1&1 Web Apps Catalog   Catalog
Category: digital media – online shops / e-commerce

Manufacturer/Client
1&1 Internet AG
Germany

Design
1&1 Internet AG
Germany

SpaceMouse Wireless   3D controller
Category: computer

Manufacturer/Client
3D Connexion
Germany

Design
Design Partners
Ireland

Filtrete™ FA-T series   Room air purifier
Category: household / tableware

Manufacturer/Client
3M Taiwan Ltd.
Taiwan

Design
PEGA D&E
Taiwan

ChocQlate   Corporate Design, packaging design
Category: food

Manufacturer/Client
4Qtrade GmbH
Germany

Design
The Hamptons Bay – Design Company
Germany

5 CUPS and some sugar   Tea packaging
Category: beverages

Manufacturer/Client
5 CUPS and some sugar
Germany

Design
Sonnenstaub – Büro für Gestaltung und Illustration
Germany

ULTIVEST   Wireless alarm system
Category: buildings

Manufacturer/Client
ABUS Security-Center GmbH & Co. KG
Germany

Design
ma design GmbH & Co. KG
Germany

Manufacturer/Client **A**

Acer K137 Projector Projector
Category: audio / video

Manufacturer/Client
Acer Inc.
Taiwan

Design
Acer Inc.
Taiwan

Acer Liquid Z5 Smartphone
Category: telecommunications

Manufacturer/Client
Acer Inc.
Taiwan

Design
Acer Inc.
Taiwan

Acer Aspire R7 Notebook
Category: computer

Manufacturer/Client
Acer Inc.
Taiwan

Design
Acer Inc.
Taiwan

Acer TravelMate P645 Notebook
Category: computer

Manufacturer/Client
Acer Inc.
Taiwan

Design
Acer Inc.
Taiwan

PEAK Notebook bag – backpack
Category: leisure / lifestyle

Manufacturer/Client
ACME Europe
Lithuania

Design
ACME Europe
Lithuania

MOON Headphones
Category: audio / video

Manufacturer/Client
ACME Europe
Lithuania

Design
ACME Europe
Lithuania

Manufacturer/Client **A**

acrylic couture — Acrylics with luxurious inlays
Category: material / textiles / wall+floor

Manufacturer/Client
Acrylic couture
Germany

Design
Acrylic couture
Germany

Casa da Copa — Product / brand experience world
Category: corporate architecture – exhibition / trade fair

Manufacturer/Client
adidas AG
Germany

Design
Lieblingsagentur GmbH
Germany
White Elements GmbH
Germany

You-Trend.Com® — Magalog: magazine catalog
Category: print media – product communication

Manufacturer/Client
admembers advertising GmbH
Germany

Design
admembers advertising GmbH
Germany

Opel Monza Concept — Concept car
Category: research+development / professional concepts

Manufacturer/Client
Adam Opel AG
Germany

Design
Adam Opel AG
Germany

Adlens® Adjustables™ — Variable focus glasses
Category: medicine / health+care

Manufacturer/Client
Adlens Ltd.
United Kingdom

Design
Adlens Ltd.
United Kingdom

THE VINCI / the-vinci.com — Protocol analyzer and converter
Category: industry / skilled trades

Manufacturer/Client
Aedilis
Lithuania

Design
ELSETA
Lithuania

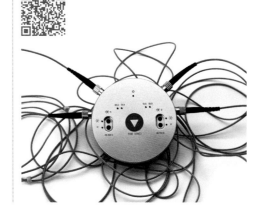

Manufacturer/Client **A**

Mom's Hurrah Detergent packaging
Category: beauty / health / household

Manufacturer/Client
Aekyung Industry
South Korea

Design
Aekyung Industry
South Korea

Springless flex frame Eyewear
Category: leisure / lifestyle

Manufacturer/Client
A & E OPTICAL Ltd.
China

Design
A & E OPTICAL Ltd.
Taiwan

NIOX VERO® Handheld medical device
Category: medicine / health+care

Manufacturer/Client
Aerocrine AB
Sweden

Design
Veryday
Sweden

Aesculap Academy Bochum Expertisium
Category: corporate architecture – shop / showroom

Manufacturer/Client
Aesculap Akademie GmbH
Germany

Design
ATELIER BRÜCKNER
Germany
LDE Belzner Holmes
Germany
Günter Hermann Architekten
Germany

LiveStore HeadSet Interactive streaming headset
Category: audio / video

Manufacturer/Client
AgênciaClick Isobar, Brazil
Fiat Automóveis S/A, Brazil

Design
Questto|Nó
Brazil

Skiwaxbrief Letter
Category: print media – advertising media

Manufacturer/Client
Agentur am Flughafen
Switzerland

Design
Agentur am Flughafen
Switzerland

Manufacturer/Client **A**

Blickwinkel   Brochure
Category: print media – corporate communication

Manufacturer/Client
Agentur am Flughafen
Switzerland

Design
Agentur am Flughafen
Switzerland

[AhnLab] MDS   Corporate security solution
Category: product interfaces

Manufacturer/Client
AhnLab
South Korea

Design
AhnLab
South Korea

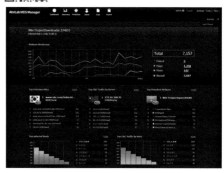

V3 25th Anniversary   Calendar and diary set
Category: print media – corporate communication

Manufacturer/Client
AhnLab
South Korea

Design
AhnLab
South Korea
Chung Choon
South Korea
Strike Communications
South Korea

JO-1MD   Color TV interphone system
Category: buildings

Manufacturer/Client
AIPHONE CO., Ltd.
Japan

Design
COBO DESIGN CO., Ltd.
Japan
AIPHONE CO., Ltd.
Japan

Bio Man   Smart shirt
Category: leisure / lifestyle

Manufacturer/Client
AiQ Smart Clothing Inc.
Taiwan

Design
AiQ Smart Clothing Inc.
Taiwan

HARD TL   Bike accessory, tire lever
Category: leisure / lifestyle

Manufacturer/Client
AIRACE ENTERPRISE Co., Ltd.
Taiwan

Design
AIRACE ENTERPRISE Co., Ltd.
Taiwan

Manufacturer/Client **A**

MINI VELOCE ROAD Bike accessory with mini pump
Category: leisure / lifestyle

Manufacturer/Client
AIRACE ENTERPRISE Co., Ltd.
Taiwan

Design
AIRACE ENTERPRISE Co., Ltd.
Taiwan

Alape A˘Form Catalog
Category: print media – product communication

Manufacturer/Client
Alape GmbH
Germany

Design
Martin et Karczinski
Germany

KeyPal Pro Bluetooth app controller
Category: telecommunications

Manufacturer/Client
Albers Inc.
Taiwan

Design
Albers Inc.
Taiwan

660336 LED wall lamp
Category: lighting

Manufacturer/Client
Albert Leuchten
Germany

Design
Albert Leuchten
Germany

662236 LED bollard lamp
Category: lighting

Manufacturer/Client
Albert Leuchten
Germany

Design
Albert Leuchten
Germany

AQUA. Mineral water Bottle and label design
Category: beverages

Manufacturer/Client
ALDI Supermercados
Spain

Design
SERIES NEMO, S. L.
Spain

Manufacturer/Client **A**

IN THE NAME OF LOVE   Exhibition catalog
Category: print media – publishing

Manufacturer/Client
Alexander Tutsek-Stiftung
Germany

Design
HUND B. communication
Germany

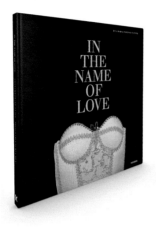

K 2.900 Silent   High-pressure cleaner
Category: leisure / lifestyle

Manufacturer/Client
Alfred Kärcher GmbH & Co. KG
Germany

Design
Alfred Kärcher GmbH & Co. KG
Germany
Teams Design GmbH
Germany

Schladerer   Corporate brochure
Category: print media – corporate communication

Manufacturer/Client
Alfred Schladerer
Germany

Design
Werbung etc.
Germany

Galke   Product catalog, calendar
Category: print media – product communication

Manufacturer/Client
Alfred Galke GmbH
Germany

Design
Heine Warnecke Design GmbH
Germany

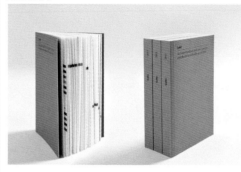

DS 6.000   Vacuum cleaner with water filter
Category: household / tableware

Manufacturer/Client
Alfred Kärcher GmbH & Co. KG
Germany

Design
Alfred Kärcher GmbH & Co. KG
Germany
BÖHLER
Corporate Industrial Design
Germany

AL-KO Drain 15000 Inox   Wastewater pump
Category: leisure / lifestyle

Manufacturer/Client
AL-KO Geräte GmbH
Germany

Design
Blankdesign
Germany
Produkt DESIGN Wolf Heieck
Germany

Manufacturer/Client **A**

AL-KO Drain 20000 HD   Sumbersible wastewater pump
Category: leisure / lifestyle

Manufacturer/Client
AL-KO Geräte GmbH
Germany

Design
Blankdesign
Germany

AL-KO Jet 4000/3 Premium   Garden pump
Category: leisure / lifestyle

Manufacturer/Client
AL-KO Geräte GmbH
Germany

Design
Blankdesign
Germany
albert ebenbichler innovations
Germany

AL-KO T20-105.4 HDE V2   Lawn tractor
Category: leisure / lifestyle

Manufacturer/Client
AL-KO Geräte GmbH
Germany

Design
Blankdesign
Germany

Allianz 1890   Customer magazine crossmedia
Category: digital media – crossmedia digital

Manufacturer/Client
Allianz Deutschland AG
Germany

Design
KircherBurkhardt GmbH
Germany

ALNO   Living book
Category: print media – product communication

Manufacturer/Client
ALNO AG
Germany

Design
Leagas Delaney Hamburg GmbH
Germany

ALNO   Cross-media campaign
Category: crossmedia – advertising / campaigns

Manufacturer/Client
ALNO AG
Germany

Design
Leagas Delaney Hamburg GmbH
Germany
VITE! Concepts GmbH
Germany

Manufacturer/Client **A**

ALNO Corporate Design
Category: crossmedia – corporate design

Manufacturer/Client
ALNO AG
Germany

Design
Leagas Delaney Hamburg GmbH
Germany

Verpackungsdesign am POS Drill bit packaging
Category: consumer electronics

Manufacturer/Client
ALPEN-MAYKESTAG GmbH
Austria

Design
ALPEN-MAYKESTAG GmbH
Austria

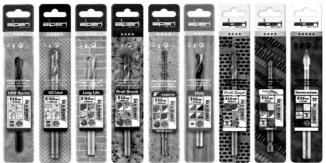

an+ sideboard Sideboard
Category: living room / bedroom

Manufacturer/Client
Alpnach
Norm-Schrankelemente AG
Switzerland

Design
dittlidesign GmbH
Switzerland

Keramo extreme Landingpage Drill bit landing page
Category: digital media – microsites

Manufacturer/Client
ALPEN-MAYKESTAG GmbH
Austria

Design
CONECTO
Business Communication GmbH
Austria

Keramo extreme Porcelain gres tile drill bit
Category: industry / skilled trades

Manufacturer/Client
ALPEN-MAYKESTAG GmbH
Austria

Design
ALPEN-MAYKESTAG GmbH
Austria

Alte Oper Frankfurt Corporate Design
Category: print media – corporate design

Manufacturer/Client
Alte Oper Frankfurt
Germany

Design
Hauser Lacour
Germany

Manufacturer/Client **A**

AMAZONE PANTERA 4502 Crop protection sprayer
Category: transportation design / special vehicles

Manufacturer/Client
AMAZONEN-Werke
H. Dreyer GmbH & Co. KG
Germany

Design
ERGO-FORM design
Germany

Admire Faucet Single lever basin mixer
Category: bathroom / wellness

Manufacturer/Client
AM.PM
Germany

Design
Gneiss Group
Denmark

Amnesty International Wailing Wall against injustice
Category: crossmedia – events

Manufacturer/Client
Amnesty International
Germany

Design
Leo Burnett GmbH
Germany

GOLD

JURY STATEMENT

"This event has been realized by Amnesty International and the artist Dan Witz from New York. They transformed building facades and walls into wailing walls for justice by inserting pictures of prison cells and windows to them. A headline is not needed because the message is all there in the images, which are very strong and really powerful. The interactive elements invite a dialogue and you feel like you can instantly do something by using the digital interfaces."

Admire WC Wallhung WC
Category: bathroom / wellness

Manufacturer/Client
AM.PM
Germany

Design
Gneiss Group
Denmark

PRIVATE SPACE Washstand
Category: bathroom / wellness

Manufacturer/Client
Anders GmbH
Germany

Design
ellenbergerdesign
Germany

Manufacturer/Client **A**

AC1200 Babyphone with monitor
Category: **leisure / lifestyle**

Manufacturer/Client
Angelcare Monitors Inc.
Canada

Design
Konings & Kappelhoff
design agency
Netherlands

SP1 NFC-enabled Bluetooth speaker
Category: **audio / video**

Manufacturer/Client
Antec, Inc.
United States of America

Design
Antec Mobile Products (A. M. P.)
United States of America

iPad Air Tablet with dramatically thinner design
Category: **computer**

Manufacturer/Client
Apple
United States of America

Design
Apple
United States of America

GOLD

JURY STATEMENT

"This iPad Air proves that it is still possible to further advance a very successful and saturated model and so it succeeds again with marvelous technology that uses even less power. A very good job was done in the power management as well as in the technology of the retina display. The iPad is lighter and thinner but the price is the same as the previous one. So all in all it has innovation, good design and a fantastic technology and operating experience."

Mac Pro Desktop computer with unified thermal core
Category: **computer**

Manufacturer/Client
Apple
United States of America

Design
Apple
United States of America

GOLD

JURY STATEMENT

"In the computer category, which has seen little evolution over the last years, Apple managed to find a completely new approach to desktop computers with a very sculptural, unique design language which derives from the functionality inside. But it is not only the design execution from the outside but also the attention to the details in the inside of the computer and the overall structure of the design which really convinced us."

Manufacturer/Client **A**

iPhone 5c   Smartphone in five colors
Category: telecommunications

Manufacturer/Client
Apple
United States of America

Design
Apple
United States of America

GOLD

JURY STATEMENT

"It is very impressive how Apple has managed to maintain their iconography despite taking a different production method with this new phone. Apple has changed the product without changing the character in itself. The new colors add freshness to the product and enliven it. This is a product that is executed very well for the market."

iPhone 5s Case   Smartphone leather case
Category: telecommunications

Manufacturer/Client
Apple
United States of America

Design
Apple
United States of America

Decostar Plus   Design room heater
Category: buildings

Manufacturer/Client
Arbonia AG
Switzerland

Design
Arbonia AG
Switzerland

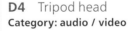

MRL4216MQ22   LTE Indoor CPE
Category: computer

Manufacturer/Client
Arcadyan Technology Corporation
Taiwan

Design
Arcadyan Technology Corporation
Taiwan

D4   Tripod head
Category: audio / video

Manufacturer/Client
ARCA-SWISS
France

Design
ARCA-SWISS
France

Manufacturer/Client **A**

BT SPEAKER   Portable audio speaker
Category: audio / video

Manufacturer/Client
Arçelik A. S.
Turkey

Design
Arçelik A. S.
Turkey

Beko Hood   Chimney hood
Category: kitchen

Manufacturer/Client
Arçelik A. S.
Turkey

Design
Arçelik A. S.
Turkey

Cast Iron   Built-in hobs
Category: kitchen

Manufacturer/Client
Arçelik A. S.
Turkey

Design
Arçelik A. S.
Turkey

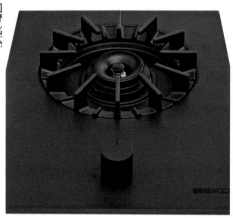

Combi Fridge   Refrigerator
Category: kitchen

Manufacturer/Client
Arçelik A. S.
Turkey

Design
Arçelik A. S.
Turkey

Grundig Side-by-Side   Refrigerator
Category: kitchen

Manufacturer/Client
Arçelik A. S.
Turkey

Design
Arçelik A. S.
Turkey

CYCLONE   Cooking system
Category: research+development / professional concepts

Manufacturer/Client
Arçelik A. S.
Turkey

Design
Arçelik A. S.
Turkey

Manufacturer/Client **A**

Grundig Slide Microwave oven
Category: research+development / professional concepts

Manufacturer/Client
Arçelik A. S.
Turkey

Design
Arçelik A. S.
Turkey

Buzz Chair
Category: living room / bedroom

Manufacturer/Client
Arco Contemporary Furniture
Netherlands

Design
Arco Contemporary Furniture
Netherlands

ABSOLUT UNIQUE Spirit bottle limited edition
Category: beverages

Manufacturer/Client
Ardagh Glass Limmared AB
Sweden

Design
FAMILY BUSINESS
Sweden

Nemo Next-generation food can
Category: food

Manufacturer/Client
Ardagh Group
Ireland

Design
Ardagh MP Germany GmbH
Germany

A³ Oven
Category: kitchen

Manufacturer/Client
Arda (Zhe Jiang) Electric Co., Ltd.
China

Design
Arda (Zhe Jiang) Electric Co., Ltd.
China

COMPACT 45 RANGE Built-in kitchen appliance
Category: kitchen

Manufacturer/Client
Arda (Zhe Jiang) Electric Co., Ltd.
China

Design
Arda (Zhe Jiang) Electric Co., Ltd.
China

Manufacturer/Client **A**

invisible neo Concealed design hinge
Category: buildings

Manufacturer/Client
Argent Alu
Belgium

Design
Argent Alu
Belgium

ZEN Spotlight
Category: lighting

Manufacturer/Client
Arkoslight
Spain

Design
Arkoslight
Spain

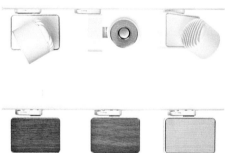

Armstrong BAU 2013 Exhibition stand
Category: corporate architecture – exhibition / trade fair

Manufacturer/Client
Armstrong DLW GmbH
Germany

Design
Ippolito Fleitz Group –
Identity Architects
Germany

Slimglot Profile lamp
Category: lighting

Manufacturer/Client
Arkoslight
Spain

Design
Arkoslight
Spain

DSA2s® MTB wheel set
Category: leisure / lifestyle

Manufacturer/Client
Armor Manufacturing Corp.
Taiwan

Design
Armor Manufacturing Corp.
Taiwan

Uni Walton Linoleum; new colors
Category: material / textiles / wall+floor

Manufacturer/Client
Armstrong DLW GmbH
Germany

Design
Armstrong DLW GmbH
Germany

Manufacturer/Client **A**

sushi 14 Magazine
Category: **print media – publishing**

Manufacturer/Client
Art Directors Club für Deutschland
(ADC) e.V.
Germany

Design
Hochschule für Gestaltung
Offenbach am Main
Germany

Chocolate LED Floor lamp
Category: **lighting**

Manufacturer/Client
Artemide S. p. A.
Italy

Design
a.g Licht
Germany

Demetra Table lamp
Category: **lighting**

Manufacturer/Client
Artemide S. p. A.
Italy

Design
NAOTO FUKASAWA DESIGN
Japan

Elle Pendant lamp
Category: **lighting**

Manufacturer/Client
Artemide S. p. A.
Italy

Design
Artemide S. p. A.
Italy

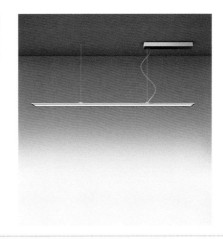

Empatia Table lamp
Category: **lighting**

Manufacturer/Client
Artemide S. p. A.
Italy

Design
Studio de Bevilacqua
Italy

Sostituto Pole top luminaire
Category: **lighting**

Manufacturer/Client
Artemide S. p. A.
Italy

Design
Artemide S. p. A.
Italy

Manufacturer/Client **A**

ARTLIFT Shower bathtub with bath lift
Category: bathroom / wellness

Manufacturer/Client
Artweger GmbH & Co. KG
Austria

Design
GP designpartners GmbH
Austria

Norco Lifestyle Bag Lifestyle bag
Category: leisure / lifestyle

Manufacturer/Client
ASISTA Teile fürs Bad
GmbH & Co. KG
Germany

Design
Kastl Design
Germany

Asstel Landingpage The worst client
Category: digital media – microsites

Manufacturer/Client
Asstel ProKunde
Germany

Design
DigitasLBi AG
Germany

GOLD

JURY STATEMENT

"Communication about insurance is typically dry and uninteresting but Asstell managed to make the landing page much more user-oriented and entertaining with this idea. One of the key strengths of this website is its ability to communicate with and entertain the person who visits it. There is a lot of creative insight behind this fun idea, which offers a tremendous benefit to the brand."

M8 Gaming PC
Category: computer

Manufacturer/Client
ASRock Incorporation
Taiwan

Design
BMW Group DesignworksUSA
Germany

ASUS S1 Series LED Projector
Category: audio / video

Manufacturer/Client
ASUSTek Computer Inc.
Taiwan

Design
ASUS Design Center
Taiwan

Manufacturer/Client **A**

Padfone mini Smartphone with tablet
Category: telecommunications

Manufacturer/Client
ASUSTek Computer Inc.
Taiwan

Design
ASUS Design Center
Taiwan

ASUS ME173 Android pad
Category: computer

Manufacturer/Client
ASUSTek Computer Inc.
Taiwan

Design
ASUS Design Center
Taiwan

ASUSPRO PU Series Laptop
Category: computer

Manufacturer/Client
ASUSTek Computer Inc.
Taiwan

Design
ASUS Design Center
Taiwan

ASUS G10AC Desktop PC
Category: computer

Manufacturer/Client
ASUSTek Computer Inc.
Taiwan

Design
ASUS Design Center
Taiwan

ASUS PA Series Professional monitor
Category: computer

Manufacturer/Client
ASUSTek Computer Inc.
Taiwan

Design
ASUS Design Center
Taiwan

ASUS VivoMouse Mouse
Category: computer

Manufacturer/Client
ASUSTek Computer Inc.
Taiwan

Design
ASUS Design Center
Taiwan

Manufacturer/Client **A**

MAXMIUS VI FORMULA Motherboard
Category: computer

Manufacturer/Client
ASUSTek Computer Inc.
Taiwan

Design
ASUS Design Center
Taiwan

ZENBOOK™ Pro Series Laptop
Category: computer

Manufacturer/Client
ASUSTek Computer Inc.
Taiwan

Design
ASUS Design Center
Taiwan

Brazilian Clichés Book
Category: print media – publishing

Manufacturer/Client
Ateliê Editorial
Brazil

Design
Casa Rex
Brazil

Transformer Book Trio Tablet / Laptop
Category: computer

Manufacturer/Client
ASUSTek Computer Inc.
Taiwan

Design
ASUS Design Center
Taiwan

ZENBOOK™ Series Laptop
Category: computer

Manufacturer/Client
ASUSTek Computer Inc.
Taiwan

Design
ASUS Design Center
Taiwan

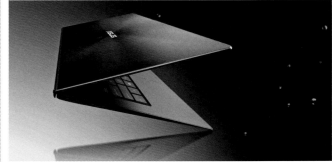

AAA Ring
Category: leisure / lifestyle

Manufacturer/Client
ATELIER ALLURE
by THOMAS HAUSER
Austria

Design
ATELIER ALLURE
by THOMAS HAUSER
Austria

Manufacturer/Client **A**

LOLA Bangle
Category: leisure / lifestyle

Manufacturer/Client
ATELIER ALLURE
by THOMAS HAUSER
Austria

Design
ATELIER ALLURE
by THOMAS HAUSER
Austria

Audi Car Configurator Tablet App
Category: digital media – mobile applications

Manufacturer/Client
AUDI AG
Germany

Design
SapientNitro
Germany

Audi A3 Limousine Vehicle
Category: transportation design / special vehicles

Manufacturer/Client
AUDI AG
Germany

Design
AUDI AG
Germany

Hood LED luminaire and lighting
Category: lighting

Manufacturer/Client
Ateljé Lyktan AB
Sweden

Design
Form Us With Love
Sweden

Audi Website for Tablet Tablet App
Category: digital media – mobile applications

Manufacturer/Client
AUDI AG
Germany

Design
SapientNitro
Germany

Audi R8 Sports car
Category: transportation design / special vehicles

Manufacturer/Client
AUDI AG
Germany

Design
AUDI AG
Germany

Manufacturer/Client **A**

Audi RS 6 Avant   Vehicle
Category: transportation design / special vehicles

Manufacturer/Client
AUDI AG
Germany

Design
AUDI AG
Germany

Audi GRC Makulaturbuch   Notebook
Category: print media – corporate design

Manufacturer/Client
AUDI AG
Germany

Design
Martin et Karczinski
Germany

AUDI AG, Paris 2012   Exhibition stand
Category: corporate architecture – exhibition / trade fair

Manufacturer/Client
AUDI AG
Germany

Design
SCHMIDHUBER / KMS BLACKSPACE
Germany

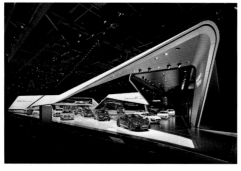

Audi RS 7 Sportback   Vehicle
Category: transportation design / special vehicles

Manufacturer/Client
AUDI AG
Germany

Design
AUDI AG
Germany

AUDI AG – Cone of Light   Exhibition stand CES 2013
Category: corporate architecture – exhibition / trade fair

Manufacturer/Client
AUDI AG
Germany

Design
tisch13 GmbH
Germany

AUGUST Smart Lock   Lock system
Category: buildings

Manufacturer/Client
AUGUST
United States of America

Design
Yves Behar – fuseproject
United States of America

Manufacturer/Client **A**

EnergyOptimizer   Product Interface
Category: product interfaces

Manufacturer/Client
AU Optronics Corporation
Taiwan

Design
Qisda Creative Design Center
Taiwan

ENTRANCE   Collection
Category: buildings

Manufacturer/Client
AUTHENTICS GmbH
Germany

Design
jehs + laub GbR
Germany

Das Brot.   Bread bakery and restaurant
Category: corporate architecture – hotel / spa / gastronomy

Manufacturer/Client
Autostadt GmbH
Germany

Design
Designliga – Büro für
Visuelle Kommunikation
und Innenarchitektur
Germany

Giro   Savings box
Category: leisure / lifestyle

Manufacturer/Client
AUTHENTICS GmbH
Germany

Design
Karen Olze
Germany

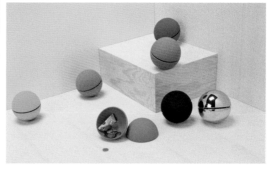

Fahrzeugkonfigurator   Showroom
Category: corporate architecture – shop / showroom

Manufacturer/Client
Autostadt GmbH
Germany

Design
jangled nerves
Germany

MobiGlobe   Interactive stations
Category: corporate architecture – communication media in architecture and public spaces

Manufacturer/Client
Autostadt GmbH
Germany

Design
Hosoya Schaefer Architects AG
Switzerland
Unity Studios
Denmark
base.io
Denmark
Shiftcontrol
Denmark
Chragokyberneticks
Switzerland
Büro Destruct
Switzerland
Idee und Klang
Switzerland
blm Filmproduktion GmbH
Germany
OPTIX Digital Pictures GmbH
Germany

Manufacturer/Client **A**

ZERO 1   3-way active loudspeaker
Category: audio / video

Manufacturer/Client
Avantgarde Acoustic GmbH
Germany

Design
Adamidesign
Germany

GOLD

JURY STATEMENT

"This speaker is a sculpture which, when imagining it in white in front of a white wall, as part of the architecture itself, is perceived as pure sound architecture – the product takes a backseat. The workmanship is excellent all around. Despite its dimensions and its extremely outstanding sound quality, the speaker as a product is utterly restrained."

myStory Box   Video collection device
Category: product interfaces

Manufacturer/Client
AVerMedia TECHNOLOGIES, Inc.
Taiwan

Design
AVerMedia TECHNOLOGIES, Inc.
Taiwan

Core Series   Gaming memory
Category: computer

Manufacturer/Client
AVEXIR Technologies Corporation
Taiwan

Design
AVEXIR Technologies Corporation
Taiwan

Explorado Kindermuseum   Corporate Design
Category: crossmedia – corporate design

Manufacturer/Client
AWC AG
Germany

Design
kleiner und bold GmbH
Germany

Backhaus Dries   Corporate Design
Category: print media – corporate design

Manufacturer/Client
Backhaus Dries GmbH
Germany

Design
Fuenfwerken Design AG
Germany

Manufacturer/Client **B**

AirEngine   Air purifier
Category: household / tableware

Manufacturer/Client
BALMUDA Inc.
Japan

Design
BALMUDA Inc.
Japan

GOLD

JURY STATEMENT

"The AirEngine air purifier impresses with its linear design and displays a fine and honest handling of the materials used. The easy replacement of filters constitutes another large plus point for the users. With its restrained design language, the product easily blends in anywhere."

SmartHeater   Heater
Category: buildings

Manufacturer/Client
BALMUDA Inc.
Japan

Design
BALMUDA Inc.
Japan

GOLD

JURY STATEMENT

"The design of the radiator is simple and unobtrusive, thought through to the very last detail – the opposite of a technically overloaded product. And in an unprecedented manner, it even integrates itself perfectly into Japanese homes. The functionality is equally convincing – the radiator is extremely efficient due to the heat conducting aluminium, it heats up quickly and is also quickly turned off."

Mistral Grand Selection   Tea packaging, food packaging
Category: packaging formdesign

Manufacturer/Client
Baliarne obchodu, a.s. Proprad
Slovakia

Design
atelier peter schmidt,
belliero & zandée
Germany

bamboo 2013   Yearbook
Category: print media – publishing

Manufacturer/Client
bamboo
Brazil

Design
estúdio lógos
Brazil

Manufacturer/Client **B**

BeoLab 14   Speaker system
Category: audio / video

Manufacturer/Client
Bang & Olufsen
Denmark

Design
Bang & Olufsen
Denmark

BeoPlay H3   Headphone
Category: audio / video

Manufacturer/Client
Bang & Olufsen
Denmark

Design
Bang & Olufsen
Denmark

BeoPlay H6   Headphone
Category: audio / video

Manufacturer/Client
Bang & Olufsen
Denmark

Design
Bang & Olufsen
Denmark

asano   Multi-room audiosystem
Category: audio / video

Manufacturer/Client
basalte
Belgium

Design
basalte
Belgium

Spinova Produktsystem   Platform for back therapy
Category: medicine / health+care

Manufacturer/Client
Bauerfeind AG
Germany

Design
Rokitta Produkt & Marken-
ästhetik
Germany

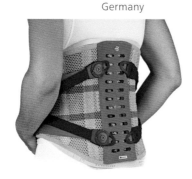

BMW Magazin App   Mobile App
Category: digital media – mobile applications

Manufacturer/Client
Bayerische Motoren Werke AG
Germany

Design
HOFFMANN UND CAMPE VERLAG
GmbH
Germany

Manufacturer/Client **B**

Haus der Berge Vertical Wilderness
Category: corporate architecture – installations in public spaces

Manufacturer/Client	Design
Bayerisches Staatsministerium Germany	ATELIER BRÜCKNER Germany TAMSCHICK MEDIA+SPACE Germany KLANGERFINDER Germany LDE Belzner Holmes Germany Staatliches Bauamt Traunstein Germany

CX2000 Embedded PC
Category: industry / skilled trades

Manufacturer/Client	Design
Beckhoff Automation GmbH Germany	design Adrian und Greiser GbR Germany

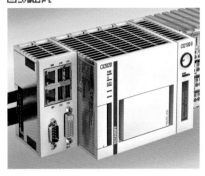

Bekina Agrilite Saftey boots
Category: industry / skilled trades

Manufacturer/Client	Design
Bekina Boots N. V. Belgium	Scherfdesign Concept & Development Germany

ODE Publication
Category: print media – publishing

Manufacturer/Client	Design
Bel Epok GmbH Germany	Bel Epok GmbH Germany

WakuWaku Dammtor Relaunch
Category: corporate architecture – hotel / spa / gastronomy

Manufacturer/Client	Design
Bella Italia Weine Germany	Ippolito Fleitz Group – Identity Architects Germany

beactive+e Walker with electric drive
Category: medicine / health+care

Manufacturer/Client	Design
BEMOTEC GmbH Germany	Tricon Design AG Germany

Manufacturer/Client **B**

RIYA Swivel chair
Category: **office / business**

Manufacturer/Client	Design
Bene AG	PearsonLloyd Design Ltd.
Austria	United Kingdom

BenQ EW series Monitor
Category: **computer**

Manufacturer/Client	Design
BenQ Corp.	BenQ Corp.
Taiwan	Taiwan

BenQ VW series Monitor
Category: **computer**

Manufacturer/Client	Design
BenQ Corp.	BenQ Corp.
Taiwan	Taiwan

BERKER GENERATION R. Campaign for product launch
Category: **crossmedia – advertising / campaigns**

Manufacturer/Client	Design
Berker GmbH & Co. KG	Thomas Biswanger Design
Germany	Germany

Bertelsmann GeschäftsberichtsApp 2012
Annual Report App
Category: **digital media – mobile applications**

Manufacturer/Client	Design
Bertelsmann SE & Co. KGaA	Bertelsmann SE & Co. KGaA
Germany	Germany

BET-Chronik Energiewende Buch
Category: **print media – corporate communication**

Manufacturer/Client	Design
BET Aachen	wesentlich. visuelle kommunikation
Germany	Germany

Manufacturer/Client **B**

Acus Office LED table lamp
Category: lighting

Manufacturer/Client
betec Licht AG
Germany

Design
betec Licht AG
Germany

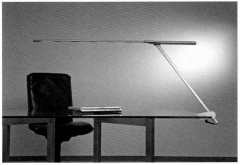

Spätzle-Blitz Kitchen appliance for Spätzle
Category: household / tableware

Manufacturer/Client
Betty Bossi AG
Switzerland

Design
Eidenbenz Industrial Design Est.
Liechtenstein

Versorgungselemente Coupling, filter, flow indicator
Category: industry / skilled trades

Manufacturer/Client
bfs·batterie füllungs systeme
GmbH
Germany

Design
bfs·batterie füllungs systeme
GmbH
Germany
Industrial Design Associates
Germany

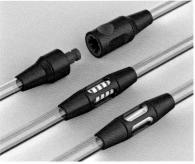

EASY HEAD ALLOY PUMP Bicycle pump
Category: leisure / lifestyle

Manufacturer/Client
BETO ENG. & MKTG. Co., Ltd.
Taiwan

Design
BETO ENG. & MKTG. Co., Ltd.
Taiwan

QLOCKTWO® LARGE CE RUST Wall clock
Category: leisure / lifestyle

Manufacturer/Client
B & F Manufacture GmbH &
Co.KG
Germany

Design
Biegert & Funk Product GmbH &
Co. KG
Germany

Volta Magazine file
Category: office / business

Manufacturer/Client
Biccateca
Brazil

Design
Pedro Paulo Venzon Filho
Brazil

Manufacturer/Client **B**

Bilfinger Brand relaunch
Category: crossmedia – corporate design

Manufacturer/Client
Bilfinger SE
Germany

Design
The Hamptons Bay –
Design Company
Germany

CD Adveniat Corporate Design
Category: print media – corporate design

Manufacturer/Client
Bischöfliche Aktion
Germany

Design
wirDesign communications AG
Germany

BlackBerry® Z10 Smartphone
Category: telecommunications

Manufacturer/Client
BlackBerry Industrial Design Team
Canada

Design
BlackBerry Industrial Design Team
Canada

Front Point Shell (Men's) Jacket
Category: material / textiles / wall+floor

Manufacturer/Client
Black Diamond Equipment Limited
United States of America

Design
Black Diamond Equipment Limited
United States of America

Hot Forge Hoody (Women's) Jacket
Category: material / textiles / wall+floor

Manufacturer/Client
Black Diamond Equipment Limited
United States of America

Design
Black Diamond Equipment Limited
United States of America

Induction Shell (Men's) Jacket
Category: material / textiles / wall+floor

Manufacturer/Client
Black Diamond Equipment Limited
United States of America

Design
Black Diamond Equipment Limited
United States of America

Manufacturer/Client **B**

Blackmagic Cinema Camera Digital cinema camera
Category: audio / video

Manufacturer/Client
Blackmagic Design
Australia

Design
Blackmagic Industrial Design Team
Australia

GOLD

JURY STATEMENT

"Here we have a professional video camera, for use with lenses with electrical aperture settings which is perfectly designed in every detail and which shows an incredible attention to material and to surface design. It is a precision instrument which exactly states its objective – the perfect optical reproduction."

Battery Converters Video format converter
Category: audio / video

Manufacturer/Client
Blackmagic Design
Australia

Design
Blackmagic Industrial Design Team
Australia

BLANCO SAGA Mixer tap
Category: kitchen

Manufacturer/Client
BLANCO GmbH & Co. KG
Germany

Design
BLANCO GmbH & Co. KG
Germany

BLANCO SELECT Waste system
Category: kitchen

Manufacturer/Client
BLANCO GmbH & Co. KG
Germany

Design
BLANCO GmbH & Co. KG
Germany

Sandwichbike Bicycle
Category: leisure / lifestyle

Manufacturer/Client
Bleijh Industrial Design
Netherlands

Design
Bleijh Industrial Design
Netherlands

Manufacturer/Client **B**

BOREA Birdfeeders
Category: leisure / lifestyle

Manufacturer/Client
blomus GmbH
Germany

Design
Augenstein Produktdesign
Germany

TEA JAY Ice tea maker
Category: household / tableware

Manufacturer/Client
blomus GmbH
Germany

Design
Flöz Industrie Design
Germany

PiOne Personal Esthetic IPL
Category: medicine / health+care

Manufacturer/Client
Bluewell Corporation
South Korea

Design
Jiyoun Kim Studio
South Korea

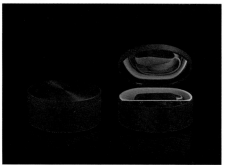

MADRA Log basket
Category: living room / bedroom

Manufacturer/Client
blomus GmbH
Germany

Design
Flöz Industrie Design
Germany

EP 300 Series Mobile Pin Pad
Category: telecommunications

Manufacturer/Client
Bluebird
South Korea

Design
Bluebird
South Korea

Blume 2000 Packaging design
Category: beauty / health / household

Manufacturer/Client
Blume 2000
Germany

Design
Studio Oeding GmbH
Germany

Manufacturer/Client **B**

Blume 2000   Corporate Design
Category: crossmedia – corporate design

Manufacturer/Client
Blume 2000
Germany

Design
Studio Oeding GmbH
Germany

Allwetterfußmatten   All-weather floor mats
Category: transportation design / special vehicles

Manufacturer/Client
BMW AG
Germany

Design
BMW AG
Germany

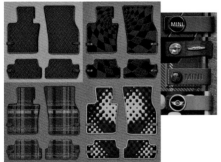

Pure – BMW i Wallbox   Wall box
Category: transportation design / special vehicles

Manufacturer/Client
BMW AG
Germany

Design
BMW Group DesignworksUSA
Germany

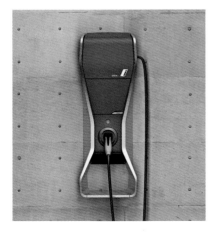

Axia 2.0   Office chair
Category: office / business

Manufacturer/Client
BMA Ergonomics
Netherlands

Design
BMA Ergonomics
Netherlands

Multifunktionsdecke   Multi-functional blanket
Category: transportation design / special vehicles

Manufacturer/Client
BMW AG
Germany

Design
BMW AG
Germany

Zubehör BMW i   BMW i accessories
Category: transportation design / special vehicles

Manufacturer/Client
BMW AG
Germany

Design
BMW AG
Germany

Manufacturer/Client **B**

MINI-Zubehör   Key cap with lanyard
Category: leisure / lifestyle

Manufacturer/Client	Design
BMW AG	BMW AG
Germany	Germany

BMW auf der IAA 2013   Exhibition concept
Category: corporate architecture – exhibition / trade fair

Manufacturer/Client	Design
BMW AG	Mutabor Design GmbH, Germany
Germany	MESO Digital Interiors GmbH
	Germany
	PIXOMONDO Studios
	GmbH & Co. KG, Germany
	Wolf Production GmbH, Germany
	yellow design, Germany

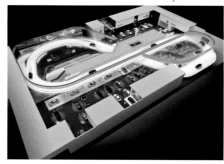

BMW i3   Automobile
Category: transportation design / special vehicles

Manufacturer/Client	Design
BMW Group	BMW Group
Germany	Germany

GOLD

JURY STATEMENT

"The BMW i3 has independent, future-orientated aesthetics and does not simply imitate vehicle models from design history. Conceptually, it is very coherent, especially with regards to how it 'grew' from the inside out – an interesting reinterpretation of 'form follows function'. What the vehicle stands for is transposed in the design."

MINI Kapooow!   Installation
Category: corporate architecture – exhibition / trade fair

Manufacturer/Client	Design
BMW Group	BMW Group
Germany	Germany

GOLD

JURY STATEMENT

"We chose the design because it simultaneously combines a vehicle and a product in a non-specialist environment. The automobile subject is not so much in the fore; it is the design process itself that is being focused on. A strong, visual image appealing to anyone with an aesthetic sense. The installation presents excellent design and preserves the iconography of the brand MINI."

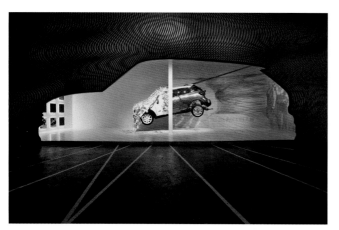

Manufacturer/Client **B**

Design Portraits. Karim Habib Film
Category: digital media – moving images

Manufacturer/Client	Design
BMW Group	BMW Group
Germany	Germany

BMW 3er Gran Turismo Automobile
Category: transportation design / special vehicles

Manufacturer/Client	Design
BMW Group	BMW Group
Germany	Germany

BMW i8 Automobile
Category: transportation design / special vehicles

Manufacturer/Client	Design
BMW Group	BMW Group
Germany	Germany

BMW 2er Coupé Automobile
Category: transportation design / special vehicles

Manufacturer/Client	Design
BMW Group	BMW Group
Germany	Germany

BMW 4er Coupé Automobile
Category: transportation design / special vehicles

Manufacturer/Client	Design
BMW Group	BMW Group
Germany	Germany

BMW R nineT Motorcycle
Category: transportation design / special vehicles

Manufacturer/Client	Design
BMW Group	BMW Group
Germany	Germany

Manufacturer/Client **B**

BMW X5 Automobile
Category: transportation design / special vehicles

Manufacturer/Client
BMW Group
Germany

Design
BMW Group
Germany

BISTRO Slicer and shredder set
Category: kitchen

Manufacturer/Client
BODUM AG
Switzerland

Design
Bodum Design Group
Switzerland

BISTRO Mixing jug
Category: household / tableware

Manufacturer/Client
BODUM AG
Switzerland

Design
Bodum Design Group
Switzerland

FYRKAT Picnic charcoal grill
Category: leisure / lifestyle

Manufacturer/Client
BODUM AG
Switzerland

Design
Bodum Design Group
Switzerland

BISTRO Meat mincer
Category: kitchen

Manufacturer/Client
BODUM AG
Switzerland

Design
Bodum Design Group
Switzerland

BISTRO Mixing bowls (5 sizes)
Category: household / tableware

Manufacturer/Client
BODUM AG
Switzerland

Design
Bodum Design Group
Switzerland

Manufacturer/Client **B**

X1 Shaving razor and stand
Category: bathroom / wellness

Manufacturer/Client
Bolin Webb Ltd.
United Kingdom

Design
Bolin Webb Ltd.
United Kingdom

BW 141 AD-5 Articulated tandem roller
Category: transportation design / special vehicles

Manufacturer/Client
BOMAG GmbH
Germany

Design
loew d.sign*
Germany

TwindexxExpress Double-deck train
Category: transportation design / special vehicles

Manufacturer/Client
Bombardier Transportation
Germany

Design
Bombardier Transportation
Germany

ZLX-12P 12" powered loudspeaker
Category: audio / video

Manufacturer/Client
Bosch Security Systems, Inc.
United States of America

Design
Teams Design
United States of America

DCN multimedia Audio powering switch
Category: audio / video

Manufacturer/Client
Bosch Sicherheitssysteme GmbH
Germany

Design
Teams Design GmbH
Germany

PLENA matrix DSP amplifier
Category: audio / video

Manufacturer/Client
Bosch Sicherheitssysteme GmbH
Germany

Design
Teams Design
China

Manufacturer/Client **B**

PLENA matrix 8 channel DSP matrix mixer
Category: audio / video

Manufacturer/Client
Bosch Sicherheitssysteme GmbH
Germany

Design
Teams Design
wChina

Schaulager Gaggenau exhibition stand
Category: corporate architecture – exhibition / trade fair

Manufacturer/Client
Bosch und Siemens
Hausgeräte GmbH
Germany

Design
eins:33 GmbH
Germany

Hair Clipper Hair clipper
Category: bathroom / wellness

Manufacturer/Client
Braun Design
Germany

Design
Braun Design
Germany

Buderus Logaflame HLS116 Heating insert
Category: buildings

Manufacturer/Client
Bosch Thermotechnik GmbH
Germany

Design
Stefan Diez
Germany

Siemens, IFA Berlin 2012 Exhibition stand
Category: corporate architecture – exhibition / trade fair

Manufacturer/Client
Bosch und Siemens
Hausgeräte GmbH
Germany

Design
SCHMIDHUBER / KMS BLACKSPACE
Germany

10 Jahre mit smart Scrapbook
Category: print media – corporate communication

Manufacturer/Client
BRAUNWAGNER GmbH
Germany

Design
BRAUNWAGNER GmbH
Germany

Manufacturer/Client **B**

Ozen Basin mixer
Category: bathroom / wellness

Manufacturer/Client
BRAVAT
Plumbing Industrial Co., Ltd.
China

Design
BRAVAT
Plumbing Industrial Co., Ltd.
China

Antibes 2 Tote bag
Category: leisure / lifestyle

Manufacturer/Client
BREE Collection GmbH & Co. KG
Germany

Design
BREE Designteam
Germany

Kiel 2 Tote bag
Category: leisure / lifestyle

Manufacturer/Client
BREE Collection GmbH & Co. KG
Germany

Design
BREE Designteam
Germany

LAPP & FAO Packaging design
Category: food

Manufacturer/Client
Bremer Feinkost GmbH & Co. KG
Germany

Design
STUDIO CHAPEAUX
Germany

Bolay & Bolay Optician store
Category: corporate architecture – shop / showroom

Manufacturer/Client
Brillengalerie Bolay & Bolay OHG
Germany

Design
LABOR WELTENBAU
Germany

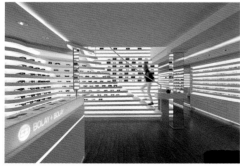

Fill&Go Water filter bottle
Category: leisure / lifestyle

Manufacturer/Client
BRITA GmbH
Germany

Design
pearl creative
Germany

Manufacturer/Client **B**

KIDFIX XP SICT   Child car seat size 2, 3
Category: leisure / lifestyle

Manufacturer/Client
Britax Römer
Kindersicherheit GmbH
Germany

Design
Britax Römer
Kindersicherheit GmbH
Germany

HL-1110/DCP-1510/MFC-1810   B/W laser printer series
Category: computer

Manufacturer/Client
Brother Industries, Ltd.
Japan

Design
Brother Industries, Ltd.
Japan

TD-2000 Series   Label printer
Category: office / business

Manufacturer/Client
Brother Industries, Ltd.
Japan

Design
Brother Industries, Ltd.
Japan

GOLD

JURY STATEMENT

"The visual simplicity of this label printer is fantastic. The product graphics are done very well. The fact that you can customize the interface in a variety of different ways shows an all-around good execution. Furthermore, it is a very iconic product and consistent in its series."

HL-3170CDW/MFC-9340CDW   Color printer series
Category: computer

Manufacturer/Client
Brother Industries, Ltd.
Japan

Design
Brother Industries, Ltd.
Japan

HL-S7000DN Series   High-speed printer
Category: computer

Manufacturer/Client
Brother Industries, Ltd.
Japan

Design
Brother Industries, Ltd.
Japan

Manufacturer/Client **B**

MFC-J6720DW/J6920DW   A3 all-in-one inkjet printer
Category: computer

Manufacturer/Client
Brother Industries, Ltd.
Japan

Design
Brother Industries, Ltd.
Japan

PT-E100   Label writer
Category: office / business

Manufacturer/Client
Brother Industries, Ltd.
Japan

Design
Brother Industries, Ltd.
Japan

A-Chair   Stacking chair
Category: office / business

Manufacturer/Client
Brunner GmbH
Germany

Design
jehs + laub GbR
Germany

Skin Labs   Visible light irradiator
Category: medicine / health+care

Manufacturer/Client
BS and Co., Ltd.
South Korea

Design
neplus
South Korea

Bosch DKS956STI   Chimney hood
Category: kitchen

Manufacturer/Client
BSH Home Appliances
(China) Co., Ltd.
China

Design
Robert Bosch Hausgeräte GmbH
Germany

Siemens H4 K28A48S0W   Refrigerator
Category: kitchen

Manufacturer/Client
BSH Home Appliances
(China) Co., Ltd.
China

Design
Siemens Electrogeräte GmbH
Germany

Manufacturer/Client **B**

Siemens KG30FS1G0C Refrigerator
Category: kitchen

Manufacturer/Client
BSH Home Appliances
(China) Co., Ltd.
China

Design
Siemens Electrogeräte GmbH
Germany

Siemens KG33NA2L0C Refrigerator
Category: kitchen

Manufacturer/Client
BSH Home Appliances
(China) Co., Ltd.
China

Design
Siemens Electrogeräte GmbH
Germany

Siemens LC45SK923W Chimney hood
Category: kitchen

Manufacturer/Client
BSH Home Appliances
(China) Co., Ltd.
China

Design
Siemens Electrogeräte GmbH
Germany

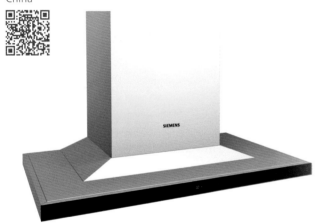

Siemens LC45SK953W Chimney hood
Category: kitchen

Manufacturer/Client
BSH Home Appliances
(China) Co., Ltd.
China

Design
Siemens Electrogeräte GmbH
Germany

Siemens LC45SK955W Chimney hood
Category: kitchen

Manufacturer/Client
BSH Home Appliances
(China) Co., Ltd.
China

Design
Siemens Electrogeräte GmbH
Germany

Siemens LS65SK950W Chimney hood
Category: kitchen

Manufacturer/Client
BSH Home Appliances
(China) Co., Ltd.
China

Design
Siemens Electrogeräte GmbH
Germany

Manufacturer/Client **B**

REST Wireless charger
Category: telecommunications

Manufacturer/Client
Bsize Inc.
Japan

Design
Bsize Inc.
Japan

SSD-WAT Thunderbolt dual SSD
Category: computer

Manufacturer/Client
BUFFALO Inc.
Japan

Design
BUFFALO Inc.
Japan

Burkhardtsmaier Teaser Teaser website
Category: digital media – microsites

Manufacturer/Client
Burkhardtsmaier
Germany

Design
Panama Werbeagentur GmbH
Germany

Buddha Spirit VI Club
Category: print media – corporate design

Manufacturer/Client
Buddha Spirit club
China

Design
ZhangYangsheng design studio
China

Hightech-Strategie Microsite campaign
Category: digital media – microsites

Manufacturer/Client
Bundesministerium für Bildung
Germany

Design
A&B One Digital GmbH
Germany

GOLD

JURY STATEMENT

"Burkhardtsmaier has created a strong message with its minimalistic teaser website that focuses on the essence of sound. The user can listen to acoustic expressions of feelings such as hope, triumph, and jealousy. This simple mechanism fittingly describes what it is like to buy products from Burkhardtsmaier. It is a great joy to visit the website, keeping users engaged with the brand in a unique way."

Manufacturer/Client **B**

Plenar2 flex Cantilever model
Category: office / business

Manufacturer/Client
Bürositzmöbelfabrik Dauphin
Germany

Design
Dauphin Design-Team
Germany

HAUS 2.0 Hand blender
Category: kitchen

Manufacturer/Client
BUWON ELECTRONICS Co., Ltd.
South Korea

Design
BUWON ELECTRONICS Co., Ltd.
South Korea

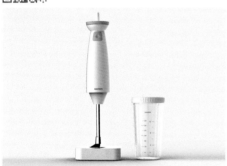

Calibre – ULTRA'GO mini Portable power station
Category: telecommunications

Manufacturer/Client
Calibre Style Ltd.
Taiwan

Design
Calibre Style Ltd.
Taiwan

Busch-ComfortPanel® 12.1 Building automation
Category: buildings

Manufacturer/Client
Busch-Jaeger Elektro GmbH
Germany

Design
ProDesign
Germany

Bowers & Wilkins P7 Headphones
Category: audio / video

Manufacturer/Client
B&W Group Ltd.
United Kingdom

Design
Native Design
United Kingdom

Calor Mini BBQ Barbeque
Category: leisure / lifestyle

Manufacturer/Client
Calor
Ireland

Design
Design Partners
Ireland

Manufacturer/Client **C**

Campus Verlag   Corporate Design
Category: crossmedia – corporate design

Manufacturer/Client
Campus Verlag GmbH
Germany

Design
Hauser Lacour
Germany

Canon Image Square   Showroom
Category: corporate architecture – shop / showroom

Manufacturer/Client
Canon Hongkong Co. Ltd.
Hong Kong

Design
Oval Design Ltd.
Hong Kong

EOS M User Interface   User Interface
Category: product interfaces

Manufacturer/Client
Canon Inc.
Japan

Design
Canon Inc.
Japan

EOS 100D   Digital camera
Category: audio / video

Manufacturer/Client
Canon Inc.
Japan

Design
Canon Inc.
Japan

EOS 5D Mark III   Digital camera
Category: audio / video

Manufacturer/Client
Canon Inc.
Japan

Design
Canon Inc.
Japan

LE-5W   Portable LED projector
Category: audio / video

Manufacturer/Client
Canon Inc.
Japan

Design
Canon Inc.
Japan

Manufacturer/Client **C**

LEGRIA mini Digital camcorder
Category: audio / video

Manufacturer/Client
Canon Inc.
Japan

Design
Canon Inc.
Japan

PowerShot N Compact digital camera
Category: audio / video

Manufacturer/Client
Canon Inc.
Japan

Design
Canon Inc.
Japan

WUX450 Projector
Category: audio / video

Manufacturer/Client
Canon Inc.
Japan

Design
Canon Inc.
Japan

i-SENSYS LBP7680Cx / 7110Cw Laser printer
Category: computer

Manufacturer/Client
Canon Inc.
Japan

Design
Canon Inc.
Japan

MG7150 / MG6450 / MG5550 All-in-one inkjet printer
Category: computer

Manufacturer/Client
Canon Inc.
Japan

Design
Canon Inc.
Japan

LA PRISE Multi-socket adaptor
Category: office / business

Manufacturer/Client
CARREFOUR
France

Design
CARREFOUR
France

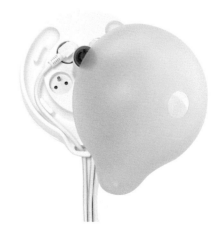

Manufacturer/Client **C**

LE MOULE   Cake tin
Category: household / tableware

Manufacturer/Client	Design
CARREFOUR	CARREFOUR
France	France

GARO-UP   Bathroom set
Category: bathroom / wellness

Manufacturer/Client	Design
Cebien Co., Ltd.	Cebien
South Korea	South Korea

LED GU10 diamond type   LED lamp
Category: lighting

Manufacturer/Client	Design
CE Lighting Ltd.	CE Lighting Ltd.
China	China

casamia 2013 calendar   Desk calendar
Category: print media – publishing

Manufacturer/Client	Design
casamia	Creative Company KKI
South Korea	South Korea

Celesio Geschäftsbericht   Annual Report 2012
Category: print media – corporate communication

Manufacturer/Client	Design
Celesio AG	Celesio AG
Germany	Germany

TV Spielfilm iPad-App   iPad App
Category: digital media – mobile applications

Manufacturer/Client	Design
CELLULAR GmbH , Germany	CELLULAR GmbH
BURDA NEWS GROUP, Germany	Germany

Manufacturer/Client **C**

ZDFheute-App Mobile App
Category: digital media – mobile applications

Manufacturer/Client
CELLULAR GmbH
Germany
ZDF, Germany

Design
CELLULAR GmbH
Germany

ZDFmediathek-App Mobile App
Category: digital media – mobile applications

Manufacturer/Client
CELLULAR GmbH
Germany
ZDF, Germany

Design
CELLULAR GmbH
Germany

Smart bedside station Medical service platform
Category: product interfaces

Manufacturer/Client
Center for Medical Informatics
South Korea

Design
pxd, Inc.
South Korea

OTTO Smart power strip
Category: office / business

Manufacturer/Client
Cerevo
Japan

Design
Satoshi Yanagisawa
Japan

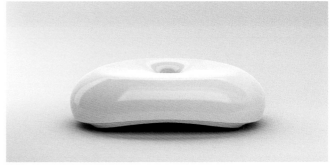

LightByte Interactive Installation
Category: research+development /professional concepts

Manufacturer/Client
Changing Places Group
United States of America

Design
MIT Media Lab
United States of America

MDLab Corporate Design
Category: crossmedia – corporate design

Manufacturer/Client
Cheil Germany GmbH
Germany

Design
Studio Oeding GmbH
Germany

Manufacturer/Client **C**

The Power Curve — Sales club
Category: corporate architecture – architecture / interior design

Manufacturer/Client
ChenFangXiao
Design International
China

Design
ChenFangXiao
Design International
China

Our life is unreal — Print media
Category: print media – advertising media

Manufacturer/Client
CHEN YU-MING
Taiwan

Design
CHEN YU-MING
Taiwan

50s news-gift paper — Posters
Category: print media – advertising media

Manufacturer/Client
Chinatown Business Association
Singapore

Design
Nanyang Technological University
Singapore

XPS300 — Laser pointer
Category: telecommunications

Manufacturer/Client
ChoisTechnology Co., Ltd.
South Korea

Design
Design Fuze Co., Ltd.
South Korea
ChoisTechnology Co., Ltd.
South Korea

Anchovy Gift Set — Gift set packaging
Category: food

Manufacturer/Client
Chung-hea Myung-ga
South Korea

Design
design workroom brill
South Korea
pentree_calligraphy
South Korea

CHP-1290D — Water purifier
Category: kitchen

Manufacturer/Client
Chungho Nais
South Korea

Design
Goth Design
South Korea
Chungho Nais Co., Ltd.
South Korea

Manufacturer/Client **C**

Dandelion Mirror   Data visualization / interaction
Category: digital media – online / offline applications

Manufacturer/Client
Chung Hua University
Taiwan

Design
Chung Hua University
Taiwan

Top Time Office   Workplace
Category: corporate architecture – architecture / interior design

Manufacturer/Client
Cimax Design Engineering
(Hong Kong) Ltd.
China

Design
Cimax Design Engineering
(Hong Kong) Ltd.
China

MX300 G2 / MX200 G2   Video conferencing systems
Category: audio / video

Manufacturer/Client
Cisco
Norway

Design
Cisco
Norway
Frost Produkt
Norway
Work
Norway

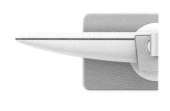

BASELWORLD CITIZEN Booth   Event booth
Category: corporate architecture – exhibition / trade fair

Manufacturer/Client
CITIZEN Watch
Japan

Design
IXI Co., Ltd.
Japan
DGT (DORELL.GHOTMEH.TANE
ARCHITECTS)
France

Beksul IceCream Mix   Ice cream powder packaging
Category: food

Manufacturer/Client
CJ CheilJedang
South Korea

Design
CJ CheilJedang
South Korea

Beksul Oil Series   Oil packaging
Category: food

Manufacturer/Client
CJ CheilJedang
South Korea

Design
CJ CheilJedang
South Korea

Manufacturer/Client **C**

Premium Oil Set Oil gift set packaging
Category: food

Manufacturer/Client
CJ CheilJedang
South Korea

Design
CJ CheilJedang
South Korea

Everygarden Superfood Vitamin packaging
Category: beauty / health / household

Manufacturer/Client
CJ CheilJedang
South Korea

Design
CJ CheilJedang
South Korea

LAMY scala Writing instruments
Category: office / business

Manufacturer/Client
C. Josef Lamy GmbH
Germany

Design
Sieger Design GmbH
Germany

CLAAS JAGUAR 800 Forage harvester
Category: transportation design / special vehicles

Manufacturer/Client
CLAAS KGaA mbH
Germany

Design
Budde Industrie Design GmbH
Germany

CLAAS AXION 800 Tractor
Category: transportation design / special vehicles

Manufacturer/Client
CLAAS KGaA mbH
Germany

Design
Budde Industrie Design GmbH
Germany

GOLD

JURY STATEMENT

"This product perfectly demonstrates how design can contribute to more quality for the user, especially regarding commercial vehicles. Improvements in cabin design, as well as an intelligent new frame structure, are greatly beneficial to the driver. The concept is captivating due to the combination of extreme performance and contemporary design."

Manufacturer/Client **C**

SCORPION 7055   Telescopic handler
Category: transportation design / special vehicles

Manufacturer/Client
CLAAS KGaA mbH
Germany

Design
Wacker Neuson Linz GmbH
Austria
Budde Industrie Design GmbH
Germany

Clever Caps 1.0   Bottle caps, building block
Category: beverages

Manufacturer/Client
Clever Pack
Brazil

Design
Clever Pack
Brazil

Clever Caps 1.0   Bottle caps, building block
Category: packaging formdesign

Manufacturer/Client
Clever Pack
Brazil

Design
Clever Pack
Brazil

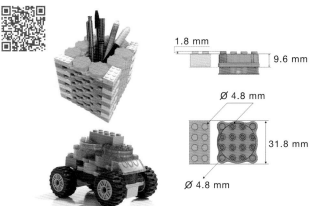

Website Clormann Design   Corporate Website
Category: digital media – corporate websites

Manufacturer/Client
Clormann Design GmbH
Germany

Design
Clormann Design GmbH
Germany

Commax Dome Camera lin-up   Video monitoring
Category: buildings

Manufacturer/Client
COMMAX
South Korea

Design
COMMAX
South Korea

hausInvest   Jubilee book
Category: print media – corporate communication

Manufacturer/Client
Commerz Real AG
Germany

Design
3st kommunikation GmbH
Germany

Manufacturer/Client **C**

KitchenMate Software App
Category: product interfaces

Manufacturer/Client
Compal Electronic, Inc.
Taiwan

Design
Compal Experience Design
Taiwan

Concorde Tablet with clever cover
Category: computer

Manufacturer/Client
Compal Electronic, Inc.
Taiwan

Design
Compal Experience Design
Taiwan

Concorde Blade Tablet with clever accessory
Category: computer

Manufacturer/Client
Compal Electronic, Inc.
Taiwan

Design
Compal Experience Design
Taiwan

Concorde Slide Tablet with clever accessory
Category: computer

Manufacturer/Client
Compal Electronic, Inc.
Taiwan

Design
Compal Experience Design
Taiwan

Duet Tablet with dual-mode dock
Category: computer

Manufacturer/Client
Compal Electronic, Inc.
Taiwan

Design
Compal Experience Design
Taiwan

Kimberley Blade edge tablet
Category: computer

Manufacturer/Client
Compal Electronic, Inc.
Taiwan

Design
Compal Experience Design
Taiwan

Manufacturer/Client **C**

Libra   Detachable laptop
Category: computer

Manufacturer/Client
Compal Electronic, Inc.
Taiwan

Design
Compal Experience Design
Taiwan

MasterChef   Kitchen accessory
Category: computer

Manufacturer/Client
Compal Electronic, Inc.
Taiwan

Design
Compal Experience Design
Taiwan

Polaris   Hybrid laptop
Category: computer

Manufacturer/Client
Compal Electronic, Inc.
Taiwan

Design
Compal Experience Design
Taiwan

Pad • NB   New notebook
Category: research+development / professional concepts

Manufacturer/Client
Compal Electronic, Inc.
Taiwan

Design
Compal Experience Design
Taiwan

Uni – Ceramic   Plastic and compound materials
Category: research+development / professional concepts

Manufacturer/Client
Compal Electronic, Inc.
Taiwan

Design
Compal Experience Design
Taiwan

Yacht pool   Swimming pool
Category: bathroom / wellness

Manufacturer/Client
Compass pools Europe
Slovakia

Design
Compass pools Europe
Slovakia

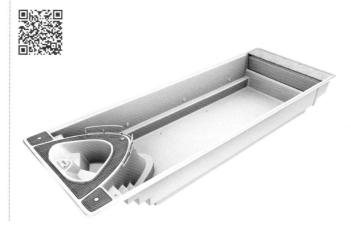

Manufacturer/Client **C**

iPin Laser Presenter — Mobile-powered laser presenter
Category: telecommunications

Manufacturer/Client
CONARY ENTERPRISE CO., Ltd.
Taiwan

Design
CONARY ENTERPRISE CO., Ltd.
Taiwan

Concord WANDERER — Travel system buggy
Category: leisure / lifestyle

Manufacturer/Client
Concord 2004, S. A.
Germany

Design
whiteID Integrated Design
Germany

Uniroyal RainSport 3 — Passenger car tire
Category: transportation design / special vehicles

Manufacturer/Client
Continental Reifen
Deutschland GmbH
Germany

Design
Continental Reifen
Deutschland GmbH
Germany

Convotherm 4 — Combi steamer
Category: kitchen

Manufacturer/Client
Convotherm Elektrogeräte GmbH
Germany

Design
Porsche Design Studio GmbH
Austria
Imago Design GmbH, Germany

Naturaline Cosmetics — Cosmetics range
Category: beauty / health / household

Manufacturer/Client
Coop Genossenschaft
Switzerland

Design
Schaffner & Conzelmann AG
Switzerland

Thread Family — Furniture series
Category: living room / bedroom

Manufacturer/Client
Coordination Ausstellungs GmbH
Germany

Design
Coordination Ausstellungs GmbH
Germany

Manufacturer/Client **C**

Giuly Design radiator
Category: buildings

Manufacturer/Client
Cordivari S. R. L.
Italy

Design
Cordivari Design
Italy

LAVIDA Luxury Time Recovery Cosmetics packaging
Category: beauty / health / household

Manufacturer/Client
Coreana Cosmetics Co., Ltd.
South Korea

Design
Coreana Cosmetics Co., Ltd.
South Korea

The Corning ONE Wireless Interactive event presentation
Category: crossmedia – events

Manufacturer/Client
Corning MobileAccess
United States of America

Design
stereolize GmbH
Germany

700K PC-Gaming-Tastatur PC gaming keyboard
Category: computer

Manufacturer/Client
COUGAR
Taiwan

Design
CRE8 DESIGN
Taiwan
COUGAR
Taiwan

700M PC-Gaming-Maus PC gaming mouse
Category: computer

Manufacturer/Client
COUGAR
Taiwan

Design
CRE8 DESIGN
Taiwan
COUGAR
Taiwan

B1 (CHPC-330N) Water purifier and coffee machine
Category: kitchen

Manufacturer/Client
COWAY
South Korea

Design
COWAY
South Korea

Manufacturer/Client **C**

CHPI-380 Water purifier and ice generator
Category: kitchen

Manufacturer/Client
COWAY
South Korea

Design
COWAY
South Korea

Juicepresso (CJP-03) Juicer
Category: kitchen

Manufacturer/Client
COWAY
South Korea

Design
COWAY
South Korea

Barrier Free Water filtration system
Category: research+development / professional concepts

Manufacturer/Client
COWAY
South Korea

Design
COWAY
South Korea

AE1 Digital car video recorder
Category: transportation design / special vehicles

Manufacturer/Client
COWON SYSTEMS Inc.
South Korea

Design
COWON SYSTEMS Inc.
South Korea

Croatian Design Review Exhibition catalog
Category: print media – publishing

Manufacturer/Client
Croatian Designers Society
Republic of Croatia

Design
Studio Sonda
Republic of Croatia

QuickPick® Remote Order picker
Category: transportation design / special vehicles

Manufacturer/Client
Crown Gabelstapler
GmbH & Co. KG
Germany

Design
Crown Gabelstapler
GmbH & Co. KG
Germany
Twisthink
United States of America

Manufacturer/Client **D**

GLSI Toolkit — Design methodology
Category: print media – information media

Manufacturer/Client
DAEHAN A&C
South Korea

Design
DAEHAN A&C
South Korea

All-In-1 Lavatory System — Bathroom
Category: bathroom / wellness

Manufacturer/Client
Daelim Dobidos
South Korea

Design
Daelim Dobidos
South Korea

Concept S-Klasse Coupé — Concept car
Category: research+development / professional concepts

Manufacturer/Client
Daimler AG
Germany

Design
Daimler AG
Germany

GOLD

JURY STATEMENT

"With the S-Class Coupe the Mercedes-Benz design team managed to focus on a clear, linear design rather than on opulence – this shows that there is a bright future ahead regarding design language. According to the jury's opinion, this positive orientation within Mercedes-Benz design deserves the iF gold award as recognition."

MB Social Publish — Website
Category: digital media – community / networking websites

Manufacturer/Client
Daimler AG
Germany

Design
Scholz & Volkmer GmbH
Germany

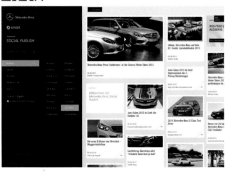

Mercedes E-Klasse — Limousine
Category: transportation design / special vehicles

Manufacturer/Client
Daimler AG
Germany

Design
Daimler AG
Germany

Manufacturer/Client **D**

Mercedes E-Klasse Cabriolet
Category: transportation design / special vehicles

Manufacturer/Client
Daimler AG
Germany

Design
Daimler AG
Germany

Mercedes S-Klasse Limousine
Category: transportation design / special vehicles

Manufacturer/Client
Daimler AG
Germany

Design
Daimler AG
Germany

CeraWall S Linear drainage-strip
Category: bathroom / wellness

Manufacturer/Client
Dallmer GmbH & Co. KG
Germany

Design
Dallmer GmbH & Co. KG
Germany

Allas Pendant Lamp
Category: research+development / professional concepts

Manufacturer/Client
Daniela Fortinho Zilinsky
Brazil

Design
Daniela Fortinho Zilinsky
Brazil

Four Treasures Of Study Stationery
Category: research+development / professional concepts

Manufacturer/Client
DAPU tech. Co., Ltd., China
HengLu Myoung Gift Co., Ltd.
China

Design
AUG design Co., Ltd.
China
DAPU tech. Co., Ltd.
China

Sonnenuhr Christmas mailing
Category: print media – advertising media

Manufacturer/Client
DART
Germany

Design
DART
Germany

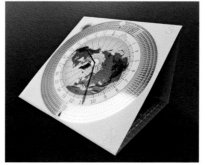

Manufacturer/Client **D**

Qixxit Mobility for a new generation
Category: product interfaces

Manufacturer/Client	Design	GOLD
DB Vertrieb GmbH	Cheil Germany GmbH	
Germany	Germany	

JURY STATEMENT

"Qixxit is a software that combines various transport carrier information in one application and thus makes compact, simple travel planning possible for the user. All information runs on a single platform. In our view, the concept is equipped with a clear, intelligent design and is also easy to operate. A complex and common problem is hereby ideally solved."

U-Boot Campaign
Category: print media – advertising media

Manufacturer/Client: DDB Tribal Wien, Austria
Design: DDB Tribal Wien, Austria

DecoSlide Adjustable wardrobe door
Category: living room / bedroom

Manufacturer/Client: DecoSlide Company A / S, Denmark
Design: Artlinco A / S, Denmark

TRAVEL BIZ Retractable cable
Category: telecommunications

Manufacturer/Client: Deff Corporation, Japan
Design: Deff Corporation, Japan

Braun MQ 5 Handblender
Category: kitchen

Manufacturer/Client: De`Longhi Braun Household GmbH, Germany
Design: Braun Design | Household, Germany

Manufacturer/Client **D**

ONE Capsule coffee machine
Category: kitchen

Manufacturer/Client
Delica AG
Switzerland

Design
2nd West
Switzerland

GOLD

JURY STATEMENT

"This coffee machine is beautiful in terms of its materials, touch, form and function. The finishing quality is also superb. There is nothing you can fault on this design, because the manufacturing quality is so high."

Bazil Exterior lighting
Category: lighting

Manufacturer/Client
DELTA LIGHT N. V.
Belgium

Design
DELTA LIGHT N. V.
Belgium

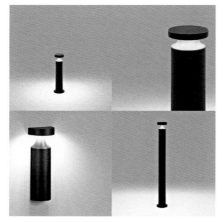

Skelp Interior lighting
Category: lighting

Manufacturer/Client
DELTA LIGHT N. V.
Belgium

Design
DELTA LIGHT N. V.
Belgium

Spotnic Interior lighting
Category: lighting

Manufacturer/Client
DELTA LIGHT N. V.
Belgium

Design
DELTA LIGHT N. V.
Belgium

Tweeter X P Exterior lighting
Category: lighting

Manufacturer/Client
DELTA LIGHT N. V.
Belgium

Design
DELTA LIGHT N. V.
Belgium

Manufacturer/Client **D**

Emergency Energy Power inverter
Category: research+development / professional concepts

Manufacturer/Client
DENSO CORPORATION
Japan

Design
DENSO CORPORATION
Japan

DAZ Limited Edition Packaging
Category: beauty / health / household

Manufacturer/Client
Design Board
Bulgaria

Design
Design Board Brussels
Belgium

Zukunftsmärkte Annual Report 2012
Category: print media – corporate communication

Manufacturer/Client
Deutsche Post DHL
Germany

Design
hw.design GmbH
Germany

WDC 2016 Taipei Bid Video Film
Category: digital media – moving images

Manufacturer/Client
Department of Cultural Affairs,
Taipei City Government
Taiwan

Design
Grass Jelly Studio
Taiwan

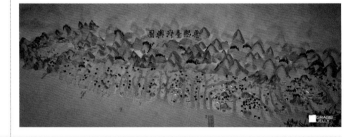

Lufthansa Mobile Services Smartphone App
Category: digital media – mobile applications

Manufacturer/Client
Deutsche Lufthansa AG
Germany

Design
SapientNitro
Germany

Volkshochschulen Corporate Design
Category: crossmedia – corporate design

Manufacturer/Client
Deutscher
Volkshochschul-Verband e.V.
Germany

Design
kleiner und bold GmbH
Germany

Manufacturer/Client **D**

Deutsches Historisches Museum   Corporate Design
Category: print media – corporate design

Manufacturer/Client
Deutsches Historisches Museum
Germany

Design
Thoma+Schekorr
Germany

Kundencenter App   Service App
Category: digital media – mobile applications

Manufacturer/Client
Deutsche Telekom AG
Germany

Design
Deutsche Telekom AG
Germany

Entertain to go   Service app
Category: digital media – crossmedia digital

Manufacturer/Client
Deutsche Telekom AG
Germany

Design
Deutsche Telekom AG
Germany

Entertain Remote Control   Service app
Category: digital media – mobile applications

Manufacturer/Client
Deutsche Telekom AG
Germany

Design
Deutsche Telekom AG
Germany

My Wallet   Digital payment app
Category: digital media – mobile applications

Manufacturer/Client
Deutsche Telekom AG
Germany

Design
Deutsche Telekom AG
Germany

2012   Annual Report
Category: print media – corporate communication

Manufacturer/Client
Deutsche Telekom AG
Germany

Design
HGB
Germany
Interbrand
Germany

Manufacturer/Client **D**

Speedphone 10 Fixed-line telephone
Category: telecommunications

Manufacturer/Client
Deutsche Telekom AG
Germany

Design
Deutsche Telekom AG
Germany

Speedport W724V Router
Category: computer

Manufacturer/Client
Deutsche Telekom AG
Germany

Design
Deutsche Telekom AG
Germany

DIE LÖSUNG Eigenwerbung Corporate Strategy
Category: print media – corporate design

Manufacturer/Client
DIE LÖSUNG – Strategie & Grafik
Germany

Design
DIE LÖSUNG – Strategie & Grafik
Germany

Im Netz der Zukunft. Brand presence IFA 2013
Category: corporate architecture – exhibition / trade fair

Manufacturer/Client
Deutsche Telekom AG
Germany

Design
q~bus Mediatektur GmbH
Germany

DEXINA AG Corporate Design
Category: print media – corporate design

Manufacturer/Client
DEXINA AG
Germany

Design
fg branddesign
Germany

LATIO Slab
Category: material / textiles / wall+floor

Manufacturer/Client
DIEPHAUS Betonwerk GmbH
Germany

Design
DIEPHAUS Betonwerk GmbH
Germany

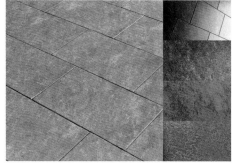

Manufacturer/Client **D**

digitalDigm's Welcome Kit Welcome package
Category: print media – corporate communication

Manufacturer/Client
digitalDigm
South Korea

Design
digitalDigm
South Korea

Patgear Portable electric scooter
Category: leisure / lifestyle

Manufacturer/Client
DIJIYA ENERGY SAVING TECH. Inc.
Taiwan

Design
DIJIYA PATGEAR
Taiwan

DCS-825L Wi-Fi baby camera
Category: leisure / lifestyle

Manufacturer/Client
D-Link Corporation
Taiwan

Design
D-Link Corporation
Taiwan

DWR-930 Mobile 4G / LTE router
Category: computer

Manufacturer/Client
D-Link Corporation
Taiwan

Design
D-Link Corporation
Taiwan

ZEIT WERT GEBEN An inspirational book containing 40 positive thoughts
Category: print media – publishing

Manufacturer/Client
dm-drogerie markt GmbH & Co. KG
Germany

Design
Projekttriangle Design Studio
Germany

GOLD

JURY STATEMENT

"The 40th anniversary of dm was the occasion to create this very inspiring, fantastic book. It is a love letter that starts from page 1 to the end held together by a nice hot pink color. The typography is very sensitive and although a whole spectrum of colors and graphic elements is used it does not look overdesigned. Every single page is wonderful and very inviting to flip through. By reading the book you will fall in love with it and so did we."

Manufacturer/Client **D**

dmSoft Typeface family
Category: print media – information media

Manufacturer/Client
dm-drogerie markt GmbH & Co. KG
Germany

Design
CLAUS KOCH™
Germany

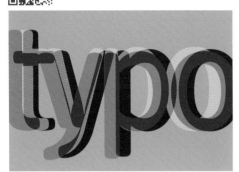

Samsung Smart Park Interactive campaign
Category: digital media – digital advertising

Manufacturer/Client
D.O.E.S
South Korea

Design
D.O.E.S
South Korea

GB v0.12 Annual Report
Category: print media – corporate communication

Manufacturer/Client
DocCheck AG
Germany

Design
antwerpes AG
Germany

GOLD

JURY STATEMENT

"This annual report is totally retro. The report is almost not designed because the design itself is limited through the ASCII fonts and the continuous-form paper. For us it is more of a collector's item than a report."

Domus Aurea Sign and guidance system
Category: corporate architecture – communication media in architecture and public spaces

Manufacturer/Client
Domus Aurea GmbH
Germany

Design
raumservice
Germany

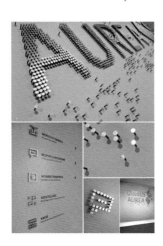

Qi Series Packaging design
Category: research+development / professional concepts

Manufacturer/Client
Dongdao Design Co., Ltd.
China

Design
Dongdao Design Co., Ltd.
China

Manufacturer/Client **D**

XS Quattro Electronic door fitting
Category: **buildings**

Manufacturer/Client
DORMA
Germany

Design
Tatic Designstudio S. R. L.
Italy

GOLD

JURY STATEMENT

"This is a very flat door fitting with integrated access control which, in an unobtrusive manner, combines every function that a door opening system, for example in the hotel sector, should feature. The sizing and application of signal colours is very reduced; on the whole, the product convinces through its unobtrusive and precise design language."

ITS 96 FL Integrated door closer
Category: **buildings**

Manufacturer/Client
DORMA
Germany

Design
DORMA
Germany

Dräger HPS 7000 Firefighter helmet
Category: **industry / skilled trades**

Manufacturer/Client
Dräger Safety AG & Co. KGaA
Germany

Design
Dräger Safety AG & Co. KGaA
Germany
Formherr Industriedesign
Germany

Dräger PARAT series Escape hood
Category: **industry / skilled trades**

Manufacturer/Client
Dräger Safety AG & Co. KGaA
Germany

Design
Dräger Safety AG & Co. KGaA
Germany
Designit Munich GmbH
Germany

911 Turbo. Die Referenz. Webspecial
Category: **digital media – microsites**

Manufacturer/Client
Dr. Ing. h.c. F. Porsche AG
Germany

Design
Bassier, Bergmann & Kindler
Germany

Manufacturer/Client **D**

Porsche – The world is a curve   Mid-engine sports car
Category: digital media – moving images

Manufacturer/Client
Dr. Ing. h.c. F. Porsche AG
Germany

Design
Kemper Kommunikation GmbH
Germany

Dropcam Pro   Wi-Fi video camera
Category: audio / video

Manufacturer/Client
Dropcam
United States of America

Design
Whipsaw Inc.
United States of America

Inipi B   Compact sauna
Category: bathroom / wellness

Manufacturer/Client
Duravit AG
Germany

Design
EOOS Design GmbH
Austria

Dr. Oetker   Corporate Website
Category: digital media – corporate websites

Manufacturer/Client
Dr. Oetker Nahrungsmittel KG
Germany

Design
SYZYGY Deutschland GmbH
Germany

DURAN® YOUTILITY   Laboratory bottle system
Category: medicine / health+care

Manufacturer/Client
DURAN Group GmbH
Germany

Design
Koop Industrial Design
Germany
Scheufele Hesse Eigler
Kommunikation, Germany
DURAN Group's technical team
Germany

GOLD

JURY STATEMENT

"Having a sauna has become more and more popular during the last 20 years; also in the private home. However, most saunas are designed for 3-4 persons. But often one has a sauna alone or as a pair and this is where the Inipi B by Duravit offers an excellent solution: attractive and intelligent detail solutions in addition to a minimal demand in space and energy provide a maximum sense of spaciousness."

Manufacturer/Client **D**

Happy D.2 Complete bathroom series
Category: **bathroom / wellness**

Manufacturer/Client
Duravit AG
Germany

Design
Sieger Design GmbH & Co. KG
Germany

Starck Badewanne Bathtub
Category: **bathroom / wellness**

Manufacturer/Client
Duravit AG
Germany

Design
Philippe Starck
France

BELLA VITA 3 Shower enclouser with sliding door
Category: **bathroom / wellness**

Manufacturer/Client
Duscholux AG
Switzerland

Design
Duscholux AG
Switzerland

SensoWash i Integrated shower-toilet
Category: **bathroom / wellness**

Manufacturer/Client
Duravit AG
Germany

Design
Philippe Starck
France

DÜRR EcoCCore Industrial cleaning system
Category: **industry / skilled trades**

Manufacturer/Client
DÜRR Ecoclean GmbH
Germany

Design
defortec GmbH
Germany

Report „Buyer's Costs" Online report
Category: **digital media – public service websites**

Manufacturer/Client
Dutch House of Representatives
Netherlands

Design
Fabrique
Netherlands

Manufacturer/Client **D**

Dyson Airblade™ Tap   Hand dryer
Category: bathroom / wellness

Manufacturer/Client
Dyson
United Kingdom

Design
Dyson
United Kingdom

Dynafly Monitor Arm   LCD monitor arm
Category: computer

Manufacturer/Client
Eastern Global Corporation
Taiwan

Design
Eastern Global Corporation
Taiwan

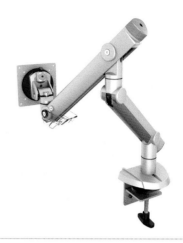

Folia™ Sugarcane Containers
Sustainable take-out container
Category: packaging formdesign

Manufacturer/Client
Eco-Products Inc.
United States of America

Design
Eco-Products Inc.
United States of America

GOLD

JURY STATEMENT

"We especially liked the aesthetics of this take-away-container. The packaging has a high-quality look to it with strong branding without being obtrusive. The clever modular concept allows for the separate accommodation of further food dishes, sauces as well as cutlery besides the main dish. This is extremely practical and presents an overall successful solution which completely convinces with its ecological aspect."

ECO OBX 20 / 18   Bearing technology
Category: buildings

Manufacturer/Client
ECO Schulte GmbH & Co. KG
Germany

Design
ECO Schulte GmbH & Co. KG
Germany

Metropole VitrA Flush   Wall-hung WC
Category: bathroom / wellness

Manufacturer/Client
Eczacibasi Yapi Gerecleri
Turkey

Design
Eczacibasi Yapi Gerecleri
Turkey

Manufacturer/Client **E**

Nest Bathroom collection
Category: bathroom / wellness

Manufacturer/Client
Eczacibasi Yapi Gerecleri
Turkey

Design
Eczacibasi Yapi Gerecleri
Turkey

Edimax EW-7438nDM Wi-Fi bridge
Category: computer

Manufacturer/Client
Edimax Technology Co., Ltd.
Taiwan

Design
Gearlab Co., Ltd.
Taiwan

Ehinger Kraftrad Corporate Design
Category: crossmedia – corporate design

Manufacturer/Client
Ehinger Kraftrad Katrin Oeding
Germany

Design
Studio Oeding GmbH
Germany

Luna Eclipse Speaker system
Category: audio / video

Manufacturer/Client
Edifier International Ltd.
Hong Kong

Design
Edifier Technology Co., Ltd.
China

CD EEW Energy from Waste Corporate Design
Category: print media – corporate design

Manufacturer/Client
EEW Energy from Waste GmbH
Germany

Design
wirDesign communications AG
Germany

BIS-KIT [DUAL PLANT POT] Plant pot and vase
Category: household / tableware

Manufacturer/Client
E. I. Corporation
South Korea

Design
E. I. Corporation
South Korea

Manufacturer/Client **E**

My Smart Growing Up Set Tableware
Category: household / tableware

Manufacturer/Client
E. I. Corporation
South Korea

Design
E. I. Corporation
South Korea

EX-G Mouse
Category: computer

Manufacturer/Client
ELECOM Co., Ltd.
Japan

Design
ELECOM Co., Ltd.
Japan

ErgoRapido Flexible cordless vacuum
Category: household / tableware

Manufacturer/Client
Electrolux AB
Sweden

Design
Electrolux Group Design
Sweden

Bluetooth Receiver Bluetooth receiver
Category: telecommunications

Manufacturer/Client
ELECOM Co., Ltd.
Japan

Design
ELECOM Co., Ltd.
Japan

Éspria Compact espresso machine
Category: kitchen

Manufacturer/Client
Electrolux AB
Sweden

Design
Electrolux Group Design
Sweden

UltraOne Vacuum cleaner
Category: household / tableware

Manufacturer/Client
Electrolux AB
Sweden

Design
Electrolux Group Design
Sweden

Manufacturer/Client **E**

QUEST E-bike configurator website
Category: digital media – online shops / e-commerce

Manufacturer/Client	Design
Elektrobiker LTD & Co. KG	studioQ
Austria	Austria

Modular LED Down light
Category: lighting

Manufacturer/Client	Design
ELR Co., Limited	ELR Co., Limited
Malaysia	Malaysia

ELV Insider Social media campaign
Category: digital media – microsites

Manufacturer/Client	Design
ELV Elektronik AG	elbkind GmbH
Germany	Germany
	Pluspol Interactive
	Germany

Intel Palace Hill Education tablet
Category: computer

Manufacturer/Client	Design
Elitegroup	Intel Corporation
Computer Systems Co., Ltd.	China
Taiwan	

System Corlo Switch series / touch display
Category: buildings

Manufacturer/Client	Design
Elsner Elektronik GmbH	Elsner Elektronik GmbH
Germany	Germany

3827 Decorative fabric
Category: material / textiles / wall+floor

Manufacturer/Client	Design
Elvin Textile Company	Elvin Textile Company
Turkey	Turkey

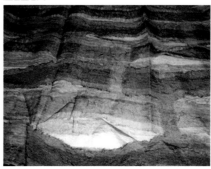

Manufacturer/Client **E**

Adjustable Wrench Ratcheting adjustable wrench
Category: industry / skilled trades

Manufacturer/Client
E-Make Tools Co., Ltd.
Taiwan

Design
E-Make Tools Co., Ltd.
Taiwan

PEACOCK Songpyeon Rice cake
Category: food

Manufacturer/Client
Emart Company Ltd.
South Korea

Design
Emart Company Ltd.
South Korea

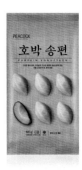

emart PLUSMATE Earphones Earphone packaging
Category: research+development / professional concepts

Manufacturer/Client
emart
South Korea

Design
emart
South Korea

GOLD

JURY STATEMENT

"The packaging directly addresses the sonic performance of the product in a way that we have never seen before. As new as they are in the category, they do not ignore the needs of good packaging: it is conceptually and communicatively very clever and graphically very humorous, furthermore it is, in contrast to other packaging, ecological due to its little use of material."

Parrish Chair
Category: public design

Manufacturer/Client
Emeco
United States of America

Design
Konstantin Grcic Industrial Design
Germany

Pling Pling Multi media player
Category: leisure / lifestyle

Manufacturer/Client
English Egg
South Korea

Design
Neolab convergence
South Korea

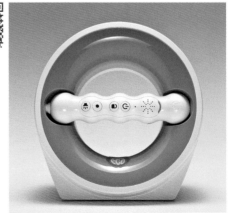

Manufacturer/Client **E**

Mythos S45 — Sauna stove
Category: bathroom / wellness

Manufacturer/Client
EOS Saunatechnik GmbH
Germany

Design
RED
Germany

Light Board — Luminaire
Category: lighting

Manufacturer/Client
ERCO GmbH
Germany

Design
ERCO GmbH
Germany

Pollux — Luminaire
Category: lighting

Manufacturer/Client
ERCO GmbH
Germany

Design
ERCO GmbH
Germany

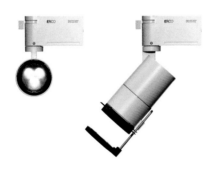

elneos — Working place system
Category: industry / skilled trades

Manufacturer/Client
erfi – Ernst Fischer GmbH & Co. KG
Germany

Design
erfi – Ernst Fischer GmbH & Co. KG
Germany

Fibermak — Laser cutting machine
Category: industry / skilled trades

Manufacturer/Client
ERMAKSAN
Turkey

Design
Arman Design and Development
Turkey

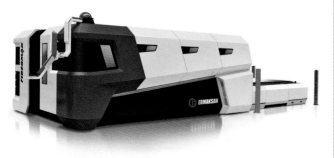

LEDA — Accent lighting (LED)
Category: lighting

Manufacturer/Client
ETAP N. V.
Belgium

Design
ALTER
France

Manufacturer/Client **E**

Eton BoostTurbine Power source
Category: telecommunications

Manufacturer/Client
Eton Corporation
United States of America

Design
Whipsaw Inc.
United States of America

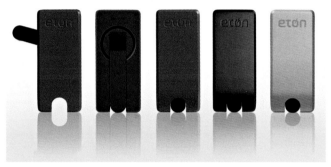

Ev. Hochschule Berlin Guidance and information system
Category: corporate architecture – communication media in architecture and public spaces

Manufacturer/Client
Evangelische Hochschule Berlin
Germany

Design
schramke design
Germany

Eva Solo To Go Grill Grill
Category: leisure / lifestyle

Manufacturer/Client
Eva Solo A / S
Denmark

Design
Eva Solo A / S
Denmark

SunLight solar power lamp Suncell driven lamp
Category: lighting

Manufacturer/Client
Eva Solo A / S
Denmark

Design
Eva Solo A / S
Denmark

Eva Solo Gravity Cookware series
Category: household / tableware

Manufacturer/Client
Eva Solo A / S
Denmark

Design
Eva Solo A / S
Denmark

Gutmann Abajo Downdraft hood
Category: kitchen

Manufacturer/Client
Exklusiv Hauben Gutmann GmbH
Germany

Design
VanBerlo B. V.
Netherlands

Manufacturer/Client **E**

Exped Packaging — Drybag packaging
Category: packaging formdesign

Manufacturer/Client
Exped AG
Switzerland

Design
Keim Identity GmbH
Switzerland

Exped Packaging — Compression bag packaging
Category: packaging formdesign

Manufacturer/Client
Exped AG
Switzerland

Design
Keim Identity GmbH
Switzerland

CLOUDY — Hanging and ceiling lamp
Category: lighting

Manufacturer/Client
Fabbian Illuminazione S. p. A.
Italy

Design
Mathieu Lehanneur since 1974
S. A. R. L
France

Ungehobene Schätze — Exhibition
Category: corporate architecture – installations in public spaces

Manufacturer/Client
facts and fiction GmbH
Germany

Design
facts and fiction GmbH
Germany

Lumina series — Controller unit
Category: industry / skilled trades

Manufacturer/Client
Fancom
Netherlands

Design
GBO DESIGN
Netherlands

Functional Backpack — Functional backpack
Category: leisure / lifestyle

Manufacturer/Client
Fashion group Hyung ji
South Korea

Design
Fashion group Hyung ji
South Korea

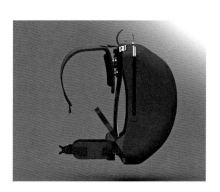

Manufacturer/Client **F**

Look inside and smile Book
Category: print media – publishing

Manufacturer/Client
FARM
Brazil

Design
FARM
Brazil

GOLD

JURY STATEMENT

"The unique celebration of the brand and their anniversary has led us to give this book an iF gold award. It allows a joyful browse through with some great three-dimensional elements to fold out or extract e.g. It is a book for inspiration as well a historical journey through the company told by its products and fabrics. The celebratory nature comes through as well as the lovely design."

Light Guide Safety Light Bike light
Category: leisure / lifestyle

Manufacturer/Client
Favour Light Company Limited
Hong Kong

Design
Favour Light Company Limited
Hong Kong

FA18 ZERO CARBON Aluminum rims
Category: leisure / lifestyle

Manufacturer/Client
FAXSON Co., Ltd.
Taiwan

Design
FAXSON Co., Ltd.
Taiwan

SMILE Audio and video door entry system
Category: buildings

Manufacturer/Client
FERMAX ELECTRONICA S. A. U.
Spain

Design
Benedito Design
Spain

F-Serie / Elektron Injection moulding machines
Category: industry / skilled trades

Manufacturer/Client
Ferromatik Milacron
Germany

Design
defortec GmbH
Germany

Manufacturer/Client **F**

NEBM Motor cable
Category: industry / skilled trades

Manufacturer/Client
Festo AG & Co. KG
Germany

Design
Festo AG & Co. KG
Germany

GOLD

JURY STATEMENT

"This motor cable is equipped with a clear, elaborate corporate identity. One just feels the attention to detail, which is evident in the choice of material, the compact shape as well as the use of colour. The shaping of the plug facilitates its handling. For us, this motor cable is an all round successful product which is why we award it the iF gold award."

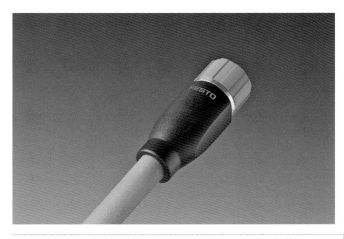

DRRD Pneumatic rotary drive
Category: industry / skilled trades

Manufacturer/Client
Festo AG & Co. KG
Germany

Design
Festo AG & Co. KG
Germany

DRVS Semi-rotary drive
Category: industry / skilled trades

Manufacturer/Client
Festo AG & Co. KG
Germany

Design
Festo AG & Co. KG
Germany

EXCM H Mini H-gantry
Category: industry / skilled trades

Manufacturer/Client
Festo AG & Co. KG
Germany

Design
Festo AG & Co. KG
Germany

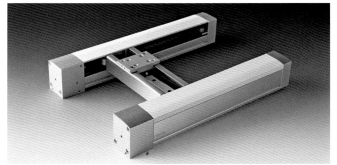

MPA-C Valve terminal
Category: industry / skilled trades

Manufacturer/Client
Festo AG & Co. KG
Germany

Design
Festo AG & Co. KG
Germany

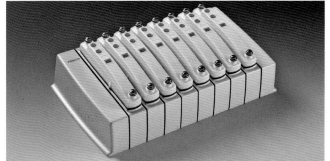

Manufacturer/Client **F**

Robotino® Mobile robot platform
Category: industry / skilled trades

Manufacturer/Client
Festo AG & Co. KG
Germany

Design
Festo AG & Co. KG
Germany

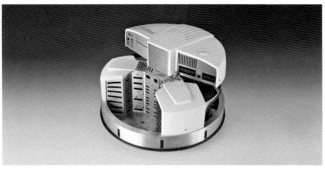

Abarth 500 – #ZeroFollowers Twitter account
Category: digital media – digital advertising

Manufacturer/Client
Fiat Group Automobiles Germany AG
Germany

Design
Leo Burnett GmbH
Germany

FIBARO Flood Sensor Detector for home systems
Category: buildings

Manufacturer/Client
Fibar Group
Poland

Design
Fibar Group
Poland

EHL 65 EQ One-handed planer
Category: industry / skilled trades

Manufacturer/Client
FESTOOL Group GmbH & Co. KG
Germany

Design
Schirrmacher Product Design
Germany

Fiat Professional Posters campaign
Category: print media – advertising media

Manufacturer/Client
Fiat Group Automobiles Germany AG
Germany

Design
Leo Burnett GmbH
Germany

Frozen Cinema Ambient Media Event
Category: crossmedia – events

Manufacturer/Client
fiftyfifty
Germany

Design
Havas Worldwide Düsseldorf
Germany

Manufacturer/Client **F**

Social Kitchen 2012 Exhibition stand
Category: corporate architecture – exhibition / trade fair

Manufacturer/Client
Fisher & Paykel
New Zealand

Design
Alt Group
New Zealand

GOLD

JURY STATEMENT

"Social Kitchen is an unusual presentation of products by a kitchen manufacturer. Clients were invited to an event in order to experience kitchen use together and then exchange views. We especially liked the pure, sensual 65 metre long staging on the theme of the basic elements earth, fire and water and which was consistently interpreted right down to the assistant's uniform."

Fitbit Force Wireless activity wristband
Category: leisure / lifestyle

Manufacturer/Client
Fitbit Inc.
United States of America

Design
NewDealDesign LLC
United States of America
Fitbit Inc.
United States of America

GOLD

JURY STATEMENT

"The reason why we thought this was an iF gold award was the comfort of this wireless activity wristband, the ergonomics were very beautiful. The design itself, the way the materials come together is incredibly accurate and precise, but it is really all about the comfort of the user. The proportion and the material intersections are handled in a very, very unique way, that puts it above all of the other products in this category at the moment."

Q! Kitchen tools
Category: household / tableware

Manufacturer/Client
Fissler GmbH
Germany

Design
Fissler GmbH
Germany
studiomem
Germany

Fleurance Nature Packaging and pop-up store
Category: beauty / health / household

Manufacturer/Client
Fleurance Nature
France

Design
The Brand Union Paris
France

Manufacturer/Client **F**

flexi VARIO   Retractable leash for dogs
Category: leisure / lifestyle

Manufacturer/Client
flexi-Bogdahn
International GmbH & Co. KG
Germany

Design
HYVE Innovation Design GmbH
Germany

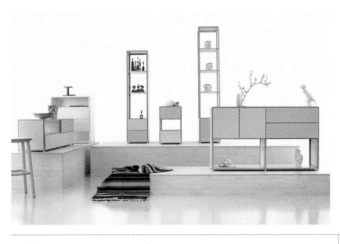

FLEYE Ori 925   Eyewear
Category: leisure / lifestyle

Manufacturer/Client
FLEYE
Denmark

Design
FLEYE
Denmark

ADD   Furniture system
Category: living room / bedroom

Manufacturer/Client
FLÖTOTTO Systemmöbel GmbH
Germany

Design
Studio Aisslinger
Germany

GOLD

JURY STATEMENT

"This furniture system is impressive with how simple and easy it is to configure with the linking element of the product. The materials used are of very high quality and beautiful. It is a cost-effective solution that still allows for a wide range of different configurations as desired. It also has an innovative linking element and a novel fastening technology for the side panels that can be attached with the push of a button."

OXeN   Heat recovery device
Category: buildings

Manufacturer/Client
FLOWAIR Sp.j.
Poland

Design
FLOWAIR Sp.j.
Poland

FLUVIA-Colección SLIM   Range of luminaires
Category: lighting

Manufacturer/Client
FLUVIA CONCEPT, S. L. U.
Spain

Design
FLUVIA CONCEPT, S. L. U.
Spain

Manufacturer/Client **F**

FBM-B 3100   Battery motor
Category: industry / skilled trades

Manufacturer/Client
FLUX-GERÄTE GmbH
Germany

Design
Teams Design GmbH
Germany

Odeon   Floor lamp
Category: lighting

Manufacturer/Client
FontanaArte S. p. A.
Italy

Design
Studio Klass
Italy

Yumi   Floor lamp
Category: lighting

Manufacturer/Client
FontanaArte S. p. A.
Italy

Design
Shigeru Ban Architects
Japan

GOLD

JURY STATEMENT

"This is a large floor lamp that is very pure and simple. It has some interesting Japanese influences and an organic expression that brings an icon into the living room. As soon as you notice the light coming from this large functional lamp, it is a surprise and feels like a little gift for the spectator. Ultimately, this is a product that is very well executed, uncomplicated and elegant."

Yupik   Portable lamp
Category: lighting

Manufacturer/Client
FontanaArte S. p. A.
Italy

Design
Form Us With Love
Sweden

OCW-01   Ohmic cooker
Category: kitchen

Manufacturer/Client
Food Industry Research and
Development Institute
Taiwan

Design
Food Industry Research and
Development Institute
Taiwan

Manufacturer/Client **F**

Ford VIGNALE   Weekender bag
Category: leisure / lifestyle

Manufacturer/Client
Ford Werke GmbH, Germany
cyber-Wear Heidelberg GmbH
Germany
Zaunkönig Lederhandwerk
Germany

Design
Ford Design | Ford Werke GmbH
Germany

Fraport AG   Signage System
Category: corporate architecture – communication media in architecture and public spaces

Manufacturer/Client
Fraport AG
Germany

Design
Hauser Lacour
Germany

Rada   Shoe rack
Category: living room / bedroom

Manufacturer/Client
Frost A / S
Denmark

Design
busk + hertzog
United Kingdom

ABEO PRO 283CGH   Photo / video tripod kit
Category: audio / video

Manufacturer/Client
Foshan Nanhai Chevan
Luxembourg

Design
Foshan Nanhai Chevan
Luxembourg

NeatConnect   Wireless scanner
Category: computer

Manufacturer/Client
frog
United States of America

Design
frog
United States of America

FUJIFILM X100S   Digital camera
Category: audio / video

Manufacturer/Client
FUJIFILM Corporation
Japan

Design
FUJIFILM Corporation
Japan

Manufacturer/Client **F**

FUJIFILM X20 Digital camera
Category: audio / video

Manufacturer/Client
FUJIFILM Corporation
Japan

Design
FUJIFILM Corporation
Japan

FUJIFILM X-M1 Digital camera
Category: audio / video

Manufacturer/Client
FUJIFILM Corporation
Japan

Design
FUJIFILM Corporation
Japan

FUJIFILM XQ1 Digital camera
Category: audio / video

Manufacturer/Client
FUJIFILM Corporation
Japan

Design
FUJIFILM Corporation
Japan

instax mini90 Instant camera
Category: audio / video

Manufacturer/Client
FUJIFILM Corporation
Japan

Design
FUJIFILM Corporation
Japan

Wonder Trip Website and special movie
Category: digital media – microsites

Manufacturer/Client
Fuji Heavy Industries Ltd.
Japan

Design
Drawing and Manual Inc.
Japan

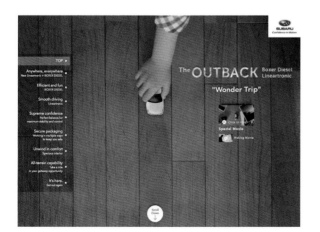

GOLD

JURY STATEMENT

"This website and film for the Outback series of Subaru is such a pleasure for the eyes from beginning to end. The storytelling is enhanced by parallax design and you feel as if you are on an actual journey that is combined with various entertaining factors. The presentation is straightforward with beautiful imagery and overall very smart."

Manufacturer/Client **F**

ARROWS NX F-01F   Smartphone
Category: telecommunications

Manufacturer/Client	**Design**
FUJITSU Ltd.	FUJITSU DESIGN Ltd.
Japan	Japan

LIFEBOOK U904   Notebook
Category: computer

Manufacturer/Client	**Design**
FUJITSU Ltd.	FUJITSU DESIGN Ltd.
Japan	Japan

Reed   Light and climate control
Category: buildings

Manufacturer/Client	**Design**
function Technology AS	function Technology AS
Norway	Norway

SMARTBALLS DUO (Serie)   Training balls
Category: leisure / lifestyle

Manufacturer/Client	**Design**
FUN FACTORY GmbH	FUN FACTORY GmbH
Germany	Germany

Revolutionär aus den Anden   Video for Gaggenau microsite
Category: digital media – moving images

Manufacturer/Client	**Design**
Gaggenau Hausgeräte GmbH	HOFFMANN UND CAMPE
Germany	VERLAG GmbH
	Germany

Gaggenau Serie 200/400   User Interface
Category: product interfaces

Manufacturer/Client	**Design**
Gaggenau Hausgeräte GmbH	Human Interface Design
Germany	Germany
	Gaggenau Hausgeräte GmbH
	Germany

Manufacturer/Client **G**

DICE+ Game controller
Category: **computer**

Manufacturer/Client
Game Technologies S. A.
Poland

Design
Mindsailors s.c.
M.Bonikowski R.Pilat
Poland

GOLD

JURY STATEMENT

"We selected these dice because of their holistic approach to bringing fun into the computer category. The element's overall good multi-functionality – you can use it as dice as well as a controller – positively surprised us. Especially notable is the well thought-through design with perfect weight distribution and the sophisticated way in which the senses are addressed. With this iF gold award, we definitely want to encourage the developers to take this approach even further."

xī xí xǐ xì Book
Category: **print media – publishing**

Manufacturer/Client
Garden City Publishers
Taiwan

Design
Wang I-Hsuan Cindy
Singapore

Midea ZL Split A / C Wall mounted air conditioner
Category: **buildings**

Manufacturer/Client
GD Midea Air-Conditioning
Equipment Co., Ltd.
China

Design
GD Midea Air-Conditioning
Equipment Co., Ltd.
China

Geberit Omega60 Actuator plate
Category: **bathroom / wellness**

Manufacturer/Client
Geberit International AG
Switzerland

Design
Christoph Behling Design Ltd.
United Kingdom

Geberit Sigma70 Actuator plate
Category: **bathroom / wellness**

Manufacturer/Client
Geberit International AG
Switzerland

Design
Christoph Behling Design Ltd.
United Kingdom

Manufacturer/Client **G**

Geberit Wandablauf   Wall drain for showers
Category: bathroom / wellness

Manufacturer/Client
Geberit International AG
Switzerland

Design
Tribecraft AG
Switzerland

Germanwings   Corporate Design
Category: print media – corporate design

Manufacturer/Client
Germanwings GmbH
Germany

Design
KW43 BRANDDESIGN
Germany

Polytector III   Gas detection device
Category: industry / skilled trades

Manufacturer/Client
GfG
Germany

Design
Lengyel Design
Germany

5AVER   Emergency mask and flashlight
Category: public design

Manufacturer/Client
GemVax & Kael
South Korea

Design
GemVax & Kael
South Korea

Microtector III   Gas detection device
Category: industry / skilled trades

Manufacturer/Client
GfG
Germany

Design
Lengyel Design
Germany

BPS X9   Banknote processing system
Category: industry / skilled trades

Manufacturer/Client
Giesecke & Devrient GmbH
Germany

Design
N+P Industrial Design GmbH
Germany

Manufacturer/Client **G**

EQUINOX MiRacle   Bicycle tire
Category: leisure / lifestyle

Manufacturer/Client
GIGANTEX
Taiwan

Design
GIGANTEX
Taiwan

Gigaset AS405 / AS405A   Cordless phone
Category: telecommunications

Manufacturer/Client
Gigaset Communications GmbH
Germany

Design
Gigaset Communications GmbH
Germany
platinumdesign
Germany

Gigaset Maxwell   Business phone
Category: telecommunications

Manufacturer/Client
Gigaset Communications GmbH
Germany

Design
Designit Munich GmbH
Germany

YTL Silk Road Tripod   Photo tripod
Category: audio / video

Manufacturer/Client
Giotto's Ind. Inc.
Taiwan

Design
Giotto's Ind. Inc.
Taiwan

SPREADTHAT !   Butter knife
Category: household / tableware

Manufacturer/Client
GIXIA GROUP Co.
Taiwan

Design
GIXIA GROUP Co.
Taiwan

PSA-01   Sound absorber module
Category: buildings

Manufacturer/Client
GIXIA GROUP Co.
Taiwan

Design
GIXIA GROUP Co.
Taiwan

PSA – 01   Sound absorbing material
Category: material / textiles / wall+floor

Manufacturer/Client
GIXIA GROUP Co.
Taiwan

Design
GIXIA GROUP Co.
Taiwan

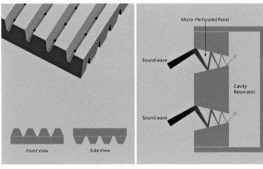

Jabra BIZ™ 2300   Corded headset
Category: audio / video

Manufacturer/Client
GN Netcom A/S
Denmark

Design
DesignIT
Denmark

Abalone gift box   Packaging design
Category: packaging formdesign

Manufacturer/Client
Goang Yuh Shing Ltd.
Taiwan

Design
Mu Creatives Co., Ltd.
Taiwan

SECUTEST   Appliance tester
Category: industry / skilled trades

Manufacturer/Client
GMC-I Messtechnik GmbH
Germany

Design
Eckstein Design
Germany

JABRA Motion UC+   Bluetooth headset for UC
Category: telecommunications

Manufacturer/Client
GN Netcom A/S
Denmark

Design
Scalae
Sweden

DYGR-3143   Vehicle protection fence
Category: public design

Manufacturer/Client
GOGANG Aluminium Co., Ltd.
South Korea

Design
GOGANG Aluminium Co., Ltd.
South Korea

Manufacturer/Client **G**

NTQ 7 Emergency stretcher
Category: medicine / health+care

Manufacturer/Client
GÖKSEL ARAS – OFİSLINE
Turkey

Design
GÖKSEL ARAS – OFİSLINE
Turkey

2nd V Pro-Balance Knife Cutlery
Category: household / tableware

Manufacturer/Client
Golden Sun Home Products Ltd.
China

Design
Golden Sun Home Products Ltd.
China

Jiminox of RAW Series Cutlery
Category: household / tableware

Manufacturer/Client
Golden Sun Home Products Ltd.
China

Design
Golden Sun Home Products Ltd.
China

Epoc Stroller
Category: leisure / lifestyle

Manufacturer/Client
Goodbaby Child Products Co., Ltd.
Netherlands

Design
GB Europe
Netherlands

Google Office Office redesign
Category: corporate architecture – architecture / interior design

Manufacturer/Client
Google Deutschland GmbH
Germany

Design
Lepel & Lepel
Germany
Jens Kirchner
Germany

HP Chromebook 11 Laptop packaging
Category: consumer electronics

Manufacturer/Client
Google, Inc.
United States of America
Hewlett-Packard Company
United States of America

Design
Google, Inc.
United States of America
Uneka
United States of America
Liquid Agency
United States of America

135

Manufacturer/Client **G**

KEF M200 Hi-fi earphones
Category: audio / video

Manufacturer/Client
GP Acoustics (UK) Ltd.
United Kingdom

Design
Porsche Design GmbH
Austria

KEF M500 Hi-fi headphones
Category: audio / video

Manufacturer/Client
GP Acoustics (UK) Ltd.
United Kingdom

Design
Porsche Design GmbH
Austria

Wetail Brochure
Category: print media – corporate communication

Manufacturer/Client
Grey Düsseldorf GmbH
Germany

Design
KW43 BRANDDESIGN
Germany

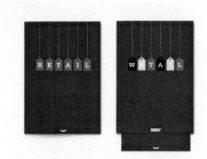

GATE Glass door lock
Category: buildings

Manufacturer/Client
Griffwerk GmbH
Germany

Design
sieger design GmbH & Co. KG
Germany

Rotella R8 Sliding door fitting
Category: buildings

Manufacturer/Client
Griffwerk GmbH
Germany

Design
Griffwerk GmbH
Germany

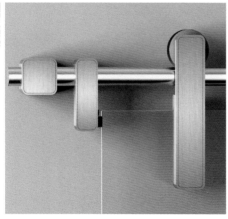

Minta Touch Kitchen faucet
Category: kitchen

Manufacturer/Client
Grohe AG
Germany

Design
Grohe AG
Germany

Manufacturer/Client **G**

D160W28MSXLRIVV-ZL(M0)   Microwave
Category: kitchen

Manufacturer/Client
Guangdong Ganlanz Microwave Oven and Electrical Appliances Manufacturing Co., Ltd., China

Design
Guangdong Ganlanz Microwave Oven and Electrical Appliances Manufacturing Co., Ltd., China

G80F25MSLVIII-ZQ(M0)   Microwave
Category: kitchen

Manufacturer/Client
Guangdong Ganlanz Microwave Oven and Electrical Appliances Manufacturing Co., Ltd., China

Design
Guangdong Ganlanz Microwave Oven and Electrical Appliances Manufacturing Co., Ltd., China

G80F25MSXLVIII-ZR(M0)   Microwave
Category: kitchen

Manufacturer/Client
Guangdong Ganlanz Microwave Oven and Electrical Appliances Manufacturing Co., Ltd., China

Design
Guangdong Ganlanz Microwave Oven and Electrical Appliances Manufacturing Co., Ltd., China

X-BOT / Colorful Pebble   Robot cleaner packaging
Category: consumer electronics

Manufacturer/Client
Guangdong Midea Kitchen
China

Design
Guangdong Midea Kitchen
China

Horizon Series   Hood and induction hob
Category: kitchen

Manufacturer/Client
Guangdong Midea Kitchen
China

Design
Guangdong Midea Kitchen
China

Find 5   Smartphone
Category: telecommunications

Manufacturer/Client
Guangdong OPPO Mobile Telecommunications Corp., Ltd.
China

Design
Guangdong OPPO Mobile Telecommunications Corp., Ltd.
China

Manufacturer/Client **G**

SeeWo Tracker   Pet tracker
Category: leisure / lifestyle

Manufacturer/Client
Guangzhou Shirui
Electronics Co., Ltd.
China

Design
Guangzhou Shiyuan
Electronic Co., Ltd.
China

Shan Shui Culture Park   Tea bag package design
Category: beverages

Manufacturer/Client
Guilin Yangshuo Nice View
China

Design
Chung Yuan Christian University
Taiwan

Times times   Office space
Category: corporate architecture – architecture / interior design

Manufacturer/Client
Guangzhou Times Property
Group
China

Design
Guangzhou Times Property
Group
China

GOLD

JURY STATEMENT

"This is a very special and unique interior of an office space of a property development company in China. This is a great representation of the company and evokes a creative image that combines art and technology. It has many different elements that work together seamlessly, including both Western and Asian influences. This is a redefined workspace that has succeeded in putting many different styles together in a very harmonious way."

Shan Shui Culture Park   Visual identity
Category: print media – corporate design

Manufacturer/Client
Guilin Yangshuo Nice View
China

Design
Chung Yuan Christian University
Taiwan
Shan Shui Branding Design Studio
Taiwan

guniPAC (RFC)   Fuel cells
Category: research+development / professional concepts

Manufacturer/Client
Gunitech Corp.
Taiwan

Design
Gunitech Corp.
Taiwan

Manufacturer/Client **G**

Caroma Marc Newson Bathroom collection
Category: bathroom / wellness

Manufacturer/Client
GWA Bathrooms & Kitchens
Australia

Design
GWA Bathrooms & Kitchens
Australia

Turntable Series TV TV
Category: audio / video

Manufacturer/Client
Haier Group
China

Design
Haier Innovation Design Center
China

Lafite Computer
Category: computer

Manufacturer/Client
Haier Group
China

Design
Industrial Design Center
of Haier Group
China

Hallenbad Ismaning Corporate Design
Category: print media – corporate communication

Manufacturer/Client
GWI Gemeindewerke Ismaning
Germany

Design
Zeichen & Wunder GmbH
Germany

Flexi All-In-One Computer
Category: computer

Manufacturer/Client
Haier Group
China

Design
Haier Innovation Design Center
China

Black King Refrigerator
Category: kitchen

Manufacturer/Client
Haier Group
China

Design
Haier Innovation Design Center
China

Manufacturer/Client **H**

Jade Refrigerator
Category: kitchen

Manufacturer/Client
Haier Group
China

Design
Haier Innovation Design Center
China

1558 Washer Washing machine
Category: household / tableware

Manufacturer/Client
Haier Group
China

Design
Haier Innovation Design Center
China

Tree in the Bottle Shampoo bottle
Category: beauty / health / household

Manufacturer/Client
Hair O'right International Corp.
Taiwan

Design
Hair O'right International Corp.
Taiwan

Hallingers Coffee labels
Category: beverages

Manufacturer/Client
Hallingers Schokoladenmanufaktur
Germany

Design
Clormann Design GmbH
Germany

Hallingers Chocolate packaging
Category: food

Manufacturer/Client
Hallingers Schokoladenmanufaktur
Germany

Design
Clormann Design GmbH
Germany

Hallingers Sortimentbox Presentation box
Category: print media – advertising media

Manufacturer/Client
Hallingers Schokoladenmanufaktur
Germany

Design
Clormann Design GmbH
Germany

Manufacturer/Client **H**

The Tea Calendar   Tea calendar
Category: print media – corporate communication

Manufacturer/Client
Hälssen & Lyon
Germany

Design
Kolle Rebbe GmbH
Germany

Art of Hangul Series   Health food packaging
Category: food

Manufacturer/Client
Hangulhwa
South Korea

Design
Hangulhwa
South Korea

da e caster   Caster
Category: living room / bedroom

Manufacturer/Client
HAMMER CASTER Co., Ltd.
Japan

Design
HOZMI DESIGN
Japan
SIMIZ Technik
Japan

GOLD

JURY STATEMENT

"This caster is beautifully executed and well thought out. Because of its elegant, unobtrusive design it harmonizes perfectly with modern furniture and fittings. The structure they used is very unconventional and the chosen materials, stainless steel and aluminium, give the caster a high quality appearance, which it definitely has."

Supreme   Kitchen appliances
Category: kitchen

Manufacturer/Client
Hangzhou ROBAM
Appliances Co., Ltd.
China

Design
Hangzhou ROBAM
Appliances Co., Ltd.
China

Nest Egg   Nest egg
Category: office / business

Manufacturer/Client
Hangzhou Teague
Technology Co., Ltd.
China

Design
Hangzhou Teague
Technology Co., Ltd.
China

Manufacturer/Client **H**

Kinergy 4s All-season performance tire
Category: transportation design / special vehicles

Manufacturer/Client
Hankook Tire Co., Ltd.
South Korea

Design
Hankook Tire Co., Ltd.
South Korea

HANSALOFT Basin mixer
Category: bathroom / wellness

Manufacturer/Client
Hansa Metallwerke AG
Germany

Design
NOA
Germany

HANSAMEDIJET FLEX Hand shower
Category: bathroom / wellness

Manufacturer/Client
Hansa Metallwerke AG
Germany

Design
NOA
Germany

HANSASTELA Basin mixer
Category: bathroom / wellness

Manufacturer/Client
Hansa Metallwerke AG
Germany

Design
NOA
Germany

AXOR-Tasche Presentation case
Category: beauty / health / household

Manufacturer/Client
Hansgrohe SE
Germany

Design
Ippolito Fleitz Group –
Identity Architects
Germany

Hansgrohe Metris 225 Basin mixer
Category: bathroom / wellness

Manufacturer/Client
Hansgrohe SE
Germany

Design
Phoenix Design GmbH & Co. KG
Germany

Manufacturer/Client **H**

Hansgrohe ShowerSelect Thermostats
Category: bathroom / wellness

Manufacturer/Client
Hansgrohe SE
Germany

Design
Phoenix Design GmbH & Co. KG
Germany

Raindance Select S 120 3jet Hand shower
Category: bathroom / wellness

Manufacturer/Client
Hansgrohe SE
Germany

Design
Phoenix Design GmbH & Co. KG
Germany

ShowerTablet Select 300 Thermostat
Category: bathroom / wellness

Manufacturer/Client
Hansgrohe SE
Germany

Design
Phoenix Design GmbH & Co. KG
Germany

MIDAS TRAY Desk tray
Category: office / business

Manufacturer/Client
HAPPY IN COMPANY
South Korea

Design
HAPPY IN COMPANY
South Korea

Filter-in-Bottle Kitchen tool
Category: household / tableware

Manufacturer/Client
HARIO Co., Ltd.
Japan

Design
HARIO Co., Ltd.
Japan

AKG K845 BT Bluetooth headphone
Category: audio / video

Manufacturer/Client
Harman International
Industries, Inc.
United States of America

Design
Harman Design Center
China

Manufacturer/Client **H**

Harman Kardon BDS 280 / 580   3D all-in-one system
Category: audio / video

Manufacturer/Client
Harman International
Industries, Inc.
United States of America

Design
Harman Design Center
China

Harman Kardon Esquire   Portable speaker
Category: audio / video

Manufacturer/Client
Harman International
Industries, Inc.
United States of America

Design
Harman Design Center
China

Harman Kardon Nova   Wireless stereo speaker
Category: audio / video

Manufacturer/Client
Harman International
Industries, Inc.
United States of America

Design
Harman Design Center
China

Harman Kardon Sabre SB35   Soundbar
Category: audio / video

Manufacturer/Client
Harman International
Industries, Inc.
United States of America

Design
Harman Design Center
China

Harman Kardon Soho   Headphones
Category: audio / video

Manufacturer/Client
Harman International
Industries, Inc.
United States of America

Design
Harman Design Center
China

Infinity Reference Series   Loudspeaker
Category: audio / video

Manufacturer/Client
Harman International
Industries, Inc.
United States of America

Design
Ashcraft Design
United States of America

Manufacturer/Client **H**

JBL E-series — Headphones
Category: audio / video

Manufacturer/Client
Harman International
Industries, Inc.
United States of America

Design
Harman Design Center
China

JBL J46BT — Wireless Bluetooth headphone
Category: audio / video

Manufacturer/Client
Harman International
Industries, Inc.
United States of America

Design
Harman International
Industries, Inc.
United States of America

JBL Spark — Home wireless speaker
Category: audio / video

Manufacturer/Client
Harman International
Industries, Inc.
United States of America

Design
Harman Design Center
China

JBL Synchros Series — Stereo headphones
Category: audio / video

Manufacturer/Client
Harman International
Industries, Inc.
United States of America

Design
Harman Design Center
China

JBL Voyager — Home wireless speaker
Category: audio / video

Manufacturer/Client
Harman International
Industries, Inc.
United States of America

Design
Harman Design Center
China

JBL Packaging — Branded packaging
Category: consumer electronics

Manufacturer/Client
Harman International
Industries, Inc.
United States of America

Design
Harman Design Center
China

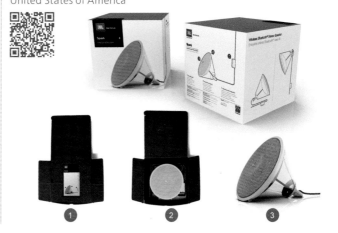

Manufacturer/Client **H**

His Master's Voice Exhibition poster
Category: print media – advertising media

Manufacturer/Client
Hartware MedienKunstVerein
Germany

Design
labor b designbüro
Germany

METAPHYS lucano Step stool and step ladder
Category: household / tableware

Manufacturer/Client
Hasegawa Kogyo
Japan

Design
hers design Inc.
Japan

Hausmann's Restaurant Corporate Design
Category: print media – corporate design

Manufacturer/Client
Hausmann's
Flughafen Gastronomie
Germany

Design
weissraum.de(sign)°
Germany

ASHIGARU Step ladder
Category: leisure / lifestyle

Manufacturer/Client
Hasegawa Kogyo
Japan

Design
DEPRO
INTERNATIONAL ASSOCIATES
Japan

Lexware LEXikon Book
Category: print media – publishing

Manufacturer/Client
Haufe-Lexware GmbH & Co. KG
Germany

Design
REINSCLASSEN GmbH & Co. KG
Germany

MeetYou Room-in-room-system
Category: office / business

Manufacturer/Client
Haworth GmbH
Germany

Design
code2design
Germany

Manufacturer/Client **H**

ROTA-S flex Manual lathe chuck
Category: industry / skilled trades

Manufacturer/Client
H.-D. SCHUNK GmbH & Co.
Germany

Design
H.-D. SCHUNK GmbH & Co.
Germany

BT Movit N-series Towing tractor / order picker
Category: transportation design / special vehicles

Manufacturer/Client
Helge Nyberg AB
Sweden

Design
Toyota Material Handling Europe AB
Sweden

Xiaomi Piston earphone Wired earphone
Category: audio / video

Manufacturer/Client
Hengyang 1more Electronic Technology Co., Ltd.
China

Design
Hengyang 1more Electronic Technology Co., Ltd.
China

Heine/Lenz/Zizka Website
Category: digital media – corporate websites

Manufacturer/Client
Heine/Lenz/Zizka Projekte GmbH
Germany

Design
Heine/Lenz/Zizka Projekte GmbH
Germany

Hello Grußkarten Stempel Greeting card-stamp set
Category: print media – advertising media

Manufacturer/Client
Hello AG
Germany

Design
Hello AG
Germany

Stiesing Magazin Magazine
Category: print media – advertising media

Manufacturer/Client
HERM. STIESING KG BREMEN
Germany

Design
POLARWERK GmbH
Germany

Manufacturer/Client **H**

Formwork   Desk accessories
Category: office / business

Manufacturer/Client
Herman Miller
United Kingdom

Design
Industrial facility
United Kingdom

GOLD

JURY STATEMENT

"This desk container is an impressive and well-thought-out solution. It is a system that offers a lot of combinations. With this system, there is a way to fulfill every demand you have in your office. The design is both simple and iconic."

Push to open Silent   Innovative opening function
Category: kitchen

Manufacturer/Client
Hettich Marketing- und Vertriebs
Germany

Design
Paul Hettich GmbH & Co. KG
Germany

Sensys   Soft closing thick door hinge
Category: kitchen

Manufacturer/Client
Hettich Marketing- und Vertriebs
Germany

Design
Hettich ONI GmbH & Co. KG
Germany

Sensys   Soft closing wide angle hinge
Category: kitchen

Manufacturer/Client
Hettich Marketing- und Vertriebs
Germany

Design
Hettich ONI GmbH & Co. KG
Germany

Sanitärsystem 815   Accessible system
Category: bathroom / wellness

Manufacturer/Client
HEWI Heinrich Wilke GmbH
Germany

Design
Phoenix Design GmbH & Co. KG
Germany

Manufacturer/Client **H**

2013 HP Low-Range Family Printer
Category: computer

Manufacturer/Client
Hewlett-Packard
United States of America

Design
Hewlett-Packard
United States of America

ENVY Recline23 TouchSmart All-in-one computer
Category: computer

Manufacturer/Client
Hewlett-Packard
United States of America

Design
Hewlett-Packard
United States of America

HP Chromebook 14 Laptop
Category: computer

Manufacturer/Client
Hewlett-Packard
United States of America

Design
Hewlett-Packard
United States of America

HP ElitePad Slate
Category: computer

Manufacturer/Client
Hewlett-Packard
United States of America

Design
Hewlett-Packard
United States of America
Native Design Ltd.
United Kingdom

HP Slatebook X2 Detachable (notebook and tablet)
Category: computer

Manufacturer/Client
Hewlett-Packard
United States of America

Design
Hewlett-Packard
United States of America

HP Spectre 13 Ultrabook Premium laptop
Category: computer

Manufacturer/Client
Hewlett-Packard
United States of America

Design
Hewlett-Packard
United States of America

Manufacturer/Client **H**

HP Spectre X2   Detachable (notebook and tablet)
Category: computer

Manufacturer/Client
Hewlett-Packard
United States of America

Design
Hewlett-Packard
United States of America

HP Split X2   Detachable (notebook and tablet)
Category: computer

Manufacturer/Client
Hewlett-Packard
United States of America

Design
Hewlett-Packard
United States of America

HP Chromebook 11   Laptop
Category: computer

Manufacturer/Client
Hewlett-Packard Company
United States of America

Design
Google
United States of America

GOLD

JURY STATEMENT

"This laptop gives you access to a free operating system which means, it creates the possibility of computing for everybody in our markets. It is designed for a particular price point and unlike many other computers in this category it reflects a balance of materials and simplicity of design and an overall honesty of execution."

HP Z8000 Bluetooth Mouse   Mobile mouse
Category: computer

Manufacturer/Client
Hewlett-Packard
United States of America

Design
Hewlett-Packard
United States of America

HP Scitex FB10000   Industrial press
Category: industry / skilled trades

Manufacturer/Client
Hewlett-Packard Company
Israel

Design
Aran Research & Development
Israel

Manufacturer/Client **H**

Hilti TE 70-D /-ATC /-AVR   Combihammer
Category: industry / skilled trades

Manufacturer/Client
Hilti Corporation
Liechtenstein

Design
Hilti Corporation
Liechtenstein
Proform Design
Germany

GOLD

JURY STATEMENT

"The Hilti combihammer is unambiguous and precise in its design; it neatly fits the Hilti range. Once again, it was demonstrated that an optimum output may be combined with a preferably low-weight device and a visually appealing design. The combihammer is a successful development; the brand is recognizable – that is perfect brand management. An absolutely contemporary product without appearing overcharged."

Hilti S-BSC   Screwdriving bit set box
Category: consumer electronics

Manufacturer/Client
Hilti Corporation
Liechtenstein

Design
Hilti Corporation
Liechtenstein
Matuschek Design & Management
Germany

Hilti S-BSP   Screwdriving bit set box
Category: consumer electronics

Manufacturer/Client
Hilti Corporation
Liechtenstein

Design
Hilti Corporation
Liechtenstein
Matuschek Design & Management
Germany

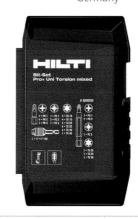

Hilti DST WRC-CA   Wireless remote control
Category: industry / skilled trades

Manufacturer/Client
Hilti Corporation
Liechtenstein

Design
Hilti Corporation
Liechtenstein
Proform Design
Germany

Hilti PD-I / PD-E   Range meters
Category: industry / skilled trades

Manufacturer/Client
Hilti Corporation
Liechtenstein

Design
Hilti Corporation
Liechtenstein
Matuschek Design & Management
Germany

Manufacturer/Client **H**

Hilti PM 4-M   Multi-line laser
Category: industry / skilled trades

Manufacturer/Client
Hilti Corporation
Liechtenstein

Design
Hilti Corporation
Liechtenstein
Matuschek Design & Management
Germany

Hilti SD-M 1   Screw magazine
Category: industry / skilled trades

Manufacturer/Client
Hilti Corporation
Liechtenstein

Design
Hilti Corporation
Liechtenstein
Matuschek Design & Management
Germany

Hilti TE 30-ATC/-AVR   Combihammer
Category: industry / skilled trades

Manufacturer/Client
Hilti Corporation
Liechtenstein

Design
Hilti Corporation
Liechtenstein
Proform Design
Germany

Hisense UX1 mobile   Mobile phone
Category: telecommunications

Manufacturer/Client
Hisense International Co., Ltd.
China

Design
Hisense Industrial Design Center
China

R-X6700D R-X6200D   Refrigerator
Category: kitchen

Manufacturer/Client
Hitachi Appliances, Inc.
Japan

Design
Hitachi, Ltd.
Japan

CG 36DAL + BL36200   Cordless grass trimmer
Category: industry / skilled trades

Manufacturer/Client
Hitachi Koki Co., Ltd.
Japan

Design
Hitachi Koki Co., Ltd.
Japan

Manufacturer/Client **H**

CH 36DL + BL36200   Cordless hedge trimmer
Category: industry / skilled trades

Manufacturer/Client
Hitachi Koki Co., Ltd.
Japan

Design
Hitachi Koki Co., Ltd.
Japan

CS 36DL + BL36200   Cordless chain saw
Category: industry / skilled trades

Manufacturer/Client
Hitachi Koki Co., Ltd.
Japan

Design
Hitachi Koki Co., Ltd.
Japan

RB 36DL + BL36200   Cordless blower
Category: industry / skilled trades

Manufacturer/Client
Hitachi Koki Co., Ltd.
Japan

Design
Hitachi Koki Co., Ltd.
Japan

CS30Y/CS35Y/CS40Y/CS45Y   Electric chain saw
Category: industry / skilled trades

Manufacturer/Client
Hitachi Koki Co., Ltd.
Japan

Design
Hitachi Koki Co., Ltd.
Japan

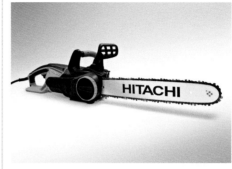

CS 51EAP   Engine chain saw
Category: industry / skilled trades

Manufacturer/Client
Hitachi Koki Co., Ltd.
Japan

Design
Hitachi Koki Co., Ltd.
Japan

PREMAX FTPi   Professional dishwasher
Category: industry / skilled trades

Manufacturer/Client
HOBART GmbH
Germany

Design
Ottenwälder und Ottenwälder
Germany

Manufacturer/Client **H**

GARANT 21 4855 — High performance milling system
Category: industry / skilled trades

Manufacturer/Client
Hoffmann GmbH Qualitäts-
werkzeuge
Germany

Design
Böhler GmbH
Germany

GARANT 65 5255_120 — Electronic torque wrench
Category: industry / skilled trades

Manufacturer/Client
Hoffmann GmbH Qualitäts-
werkzeuge
Germany

Design
Böhler Corporate Industrial
Design
Germany

GARANT 96 2480 — Industrial workchair series
Category: industry / skilled trades

Manufacturer/Client
Hoffmann GmbH Qualitäts-
werkzeuge
Germany

Design
Böhler GmbH
Germany

Honeywell Instinct — Safety spectacles
Category: industry / skilled trades

Manufacturer/Client
Honeywell Safety Products
United States of America
Honeywell Safety Products, Germany

Design
Honeywell Safety Products
United States of America

Café bord de Mer — Hotel Causal Dining
Category: corporate architecture – hotel / spa / gastronomy

Manufacturer/Client
Hong Kong Resort Company
Limited
Hong Kong

Design
Kinney Chan and Associates
Hong Kong

PG-300 Series — Portable gas analyzer
Category: industry / skilled trades

Manufacturer/Client
HORIBA, Ltd.
Japan

Design
HORIBA, Ltd.
Japan
U:GO Designers Office
Japan
Office DO
Japan

Manufacturer/Client **H**

YENGA   Design radiator
Category: buildings

Manufacturer/Client
HSK Duschkabinenbau KG
Germany

Design
HSK Duschkabinenbau KG
Germany

Lowboard   Set-Top Box
Category: audio / video

Manufacturer/Client
Huawei Device Co., Ltd.
China

Design
Design 3
Germany
Huawei Device Co., Ltd.
China

New HTC ONE   Smartphone
Category: telecommunications

Manufacturer/Client
HTC
Taiwan

Design
HTC
Taiwan

GOLD

JURY STATEMENT

"The HTC One is a well done, high-quality phone with a lot of precision in it and with a lot of effort in the detailing. It fits perfectly into your palm due to its curved backside, which offers a great product experience. It is a very well-polished phone with quality that propels HTC to new heights."

Ascend P6   Smartphone
Category: telecommunications

Manufacturer/Client
Huawei Device Co., Ltd.
China

Design
Huawei Device Co., Ltd.
China

HW-01F   Mobile Wi-Fi router
Category: telecommunications

Manufacturer/Client
Huawei Device Co., Ltd.
China

Design
Huawei Technologies Japan K. K.
Japan

Manufacturer/Client **H**

Simplicity Mobile Wi-Fi router
Category: telecommunications

Manufacturer/Client
Huawei Device Co., Ltd.
China

Design
Huawei Device Co., Ltd.
China

Aerofoil Wireless router
Category: computer

Manufacturer/Client
Huawei Device Co., Ltd.
China

Design
Huawei Device Co., Ltd.
China
Bould Design
United States of America

Aurora 4G router
Category: computer

Manufacturer/Client
Huawei Device Co., Ltd.
China

Design
Huawei Device Co., Ltd.
China

ATOM Small cell
Category: telecommunications

Manufacturer/Client
Huawei Technologies Co., Ltd.
China

Design
Huawei Technologies Co., Ltd.
China
Design 3
Germany

Tianjin Cinema Cinema
Category: corporate architecture – architecture / interior design

Manufacturer/Client
Hubei Insun Cinema Film Co., Ltd.
China

Design
One Plus Partnership Ltd.
Hong Kong

Chisels EDC Wood chisels
Category: industry / skilled trades

Manufacturer/Client
Hultafors Group AB
Sweden

Design
Hultafors Group AB
Sweden

Manufacturer/Client **H**

HumaStar 100/200   Random-access analyzer
Category: medicine / health+care

Manufacturer/Client
HUMAN
Gesellschaft für Biochemica
Germany

Design
industrialpartners GmbH
Germany

Duette® SmartCord®   Retractable cord operation
Category: buildings

Manufacturer/Client
HUNTER DOUGLAS EUROPE B. V.
Netherlands

Design
HUNTER DOUGLAS EUROPE B. V.
Netherlands

HH-SBF06   Juicer
Category: kitchen

Manufacturer/Client
Hurom L. S. Co., Ltd.
South Korea

Design
Hurom L. S. Co., Ltd.
South Korea

Duette® LiteRise®   Cordless system
Category: buildings

Manufacturer/Client
HUNTER DOUGLAS EUROPE B. V.
Netherlands

Design
HUNTER DOUGLAS EUROPE B. V.
Netherlands

HÜPPE Enjoy pure   Shower enclosure
Category: bathroom / wellness

Manufacturer/Client
HÜPPE GmbH
Germany

Design
Phoenix Design GmbH & Co. KG
Germany

HK-BBF06   Juicer
Category: kitchen

Manufacturer/Client
Hurom L. S. Co., Ltd.
South Korea

Design
CYPHICS Co., Ltd.
South Korea

Manufacturer/Client **H**

mooods   Art project #4
Category: print media – advertising media

Manufacturer/Client
hw.design GmbH
Germany

Design
hw.design GmbH
Germany

Elastic bicycle rack   Bicycle rack
Category: public design

Manufacturer/Client
Hyundai Amco
South Korea

Design
Hyundai Amco, South Korea
Spacetalk, South Korea

My Taxi   Public service design
Category: research+development /professional concepts

Manufacturer/Client
Hyundai Card
South Korea
KIA Motors Corporation
South Korea

Design
Hyundai Card
South Korea
KIA Motors Corporation
South Korea

GOLD

JURY STATEMENT

"My Taxi is a driving service which is customised to the urban conditions of the city Seoul and which focuses on the passenger's comfort. The overall concept, which addresses the prevailing problem, abolishing it holistically, is convincing. It is a clever design which grants the passenger great comfort. Every tool is thought through, from the stowage of luggage via interactive communication up to the payment system."

O-bin   Recycling bin
Category: public design

Manufacturer/Client
Hyundai Amco
South Korea

Design
Hyundai Amco, South Korea
Spacetalk, South Korea

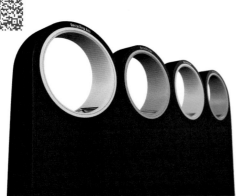

Genesis   Premium sedan
Category: transportation design / special vehicles

Manufacturer/Client
Hyundai Motor Company
South Korea

Design
Hyundai Design Center
South Korea

Manufacturer/Client **H**

FLUIDIC   Interactive light installation
Category: corporate architecture – installations in public spaces

Manufacturer/Client
Hyundai Motor Company
Germany

Design
WHITEvoid
interactive art & design
Germany

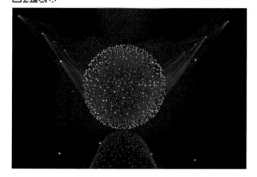

Mobius Loop   Video wall
Category: digital media – moving images

Manufacturer/Client
Hyundai Motor Group
South Korea

Design
Universal Everything
United Kingdom
KIA Design Team, Germany
Do Ho Suh, United Kingdom
INNOCEAN Worldwide
South Korea

Who Am We?   Interactive media wall
Category: corporate architecture – installations in public spaces

Manufacturer/Client
Hyundai Motor Group
South Korea

Design
Do Ho Suh
United Kingdom
Imagebakery, South Korea
Suh Architects, South Korea
INNOCEAN Worldwide
South Korea

IHN-1020ML / IDC-840SL   Home network system
Category: buildings

Manufacturer/Client
icontrols
South Korea

Design
icontrols
South Korea

BY CHOCOLATE CAFE   Café
Category: print media – corporate design

Manufacturer/Client
IDEA DO IT
South Korea

Design
IDEA DO IT
South Korea

CERAPLAN III   Fitting series
Category: bathroom / wellness

Manufacturer/Client
IDEAL STANDARD INTERNATIONAL BVBA
Belgium

Design
ARTEFAKT product design
Germany

D'LIGHT: kinetic lighting — Lighting
Category: lighting

Manufacturer/Client
ID+IM Design Lab., KAIST
South Korea

Design
ID+IM Design Lab., KAIST
South Korea

Stream Cooler — Portable cooler
Category: research+development / professional concepts

Manufacturer/Client
ID+IM Design Lab., KAIST
South Korea

Design
ID+IM Design Lab., KAIST
South Korea

3-Directional Toothbat — Dental floss holder
Category: bathroom / wellness

Manufacturer/Client
ID Infinity Limited
Hong Kong

Design
ID Infinity Limited
Hong Kong

iDuctor — Induction heating tool
Category: industry / skilled trades

Manufacturer/Client
iDtools B. V.
Netherlands

Design
FLEX/the INNOVATIONLAB B. V.
Netherlands

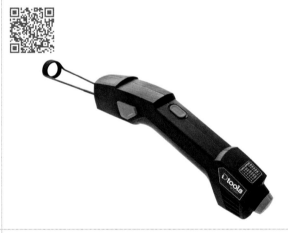

O6 wetline — Photoelectric sensor
Category: industry / skilled trades

Manufacturer/Client
ifm electronic
Germany

Design
ifm electronic
Germany

drylin® SLT — Slim linear axis
Category: industry / skilled trades

Manufacturer/Client
igus® GmbH
Germany

Design
igus® GmbH
Germany

Manufacturer/Client I

R4.1 light   Energy chain system
Category: industry / skilled trades

Manufacturer/Client
igus® GmbH
Germany

Design
igus® GmbH
Germany

readychain® Rack   Assembly transportation frame
Category: industry / skilled trades

Manufacturer/Client
igus® GmbH
Germany

Design
igus® GmbH
Germany

M2 iBaby   Baby monitor
Category: leisure / lifestyle

Manufacturer/Client
iHealth Lab Inc.
United States of America

Design
Beijing FromD Design Consultancy
China

AM3S   Activity monitor
Category: medicine / health+care

Manufacturer/Client
iHealth Lab Inc.
United States of America

Design
Beijing FromD Design Consultancy
China

BG1   Glucose meter
Category: medicine / health+care

Manufacturer/Client
iHealth Lab Inc.
United States of America

Design
Beijing FromD Design Consultancy
China

BG5   Glucose meter
Category: medicine / health+care

Manufacturer/Client
iHealth Lab Inc.
United States of America

Design
Beijing FromD Design Consultancy
China

Manufacturer/Client **I**

BP5 Wireless blood pressure monitor
Category: medicine / health+care

Manufacturer/Client
iHealth Lab Inc.
United States of America

Design
Beijing FromD Design Consultancy
China

BP7 Blood pressure monitoring device
Category: medicine / health+care

Manufacturer/Client
iHealth Lab Inc.
United States of America

Design
Beijing FromD Design Consultancy
China

HS4 Smart scale
Category: medicine / health+care

Manufacturer/Client
iHealth Lab Inc.
United States of America

Design
Beijing FromD Design Consultancy
China

PO3 Oximeter
Category: medicine / health+care

Manufacturer/Client
iHealth Lab Inc.
United States of America

Design
Beijing FromD Design Consultancy
China

Leimu Lamp
Category: lighting

Manufacturer/Client
Iittala Group Oy Ab
Finland

Design
Magnus Pettersen
United Kingdom

GOLD

JURY STATEMENT

"This lamp impresses with its usage of three different materials that are combined in a very authentic way to make an object that can brighten the domestic environment beautifully. These long-lasting materials provide a very gentle light that diffuses nicely and feels as if it is glowing all the time."

Manufacturer/Client

Vakka Box
Category: living room / bedroom

Manufacturer/Client
Iittala Group Oy Ab
Finland

Design
Aalto + Aalto
Finland

GOLD

JURY STATEMENT

"Just the simplicity and proportions of this storage system are so well executed. The ability to swap, match and stack is great. There is nothing you could fault about the production or the quality. It is so simple yet beautiful."

NH4000 High pressure distribution system
Category: industry / skilled trades

Manufacturer/Client
ILSHIN AUTOCLAVE
South Korea

Design
Tangerine & Partners
South Korea

NLA300 High pressure distribution system
Category: industry / skilled trades

Manufacturer/Client
ILSHIN AUTOCLAVE
South Korea

Design
Tangerine & Partners
South Korea

ego Bicycle lamp
Category: leisure / lifestyle

Manufacturer/Client
ilusyd
South Korea

Design
ilusyd
South Korea

Impossible Instant Film Intant film packaging
Category: consumer electronics

Manufacturer/Client
Impossible
Austria

Design
Heine/Lenz/Zizka Projekte GmbH
Germany

Manufacturer/Client I

perfect 1 second   Tent
Category: leisure / lifestyle

Manufacturer/Client
INDGROUP
South Korea

Design
INDGROUP
South Korea

Aquarius   LED light bulb
Category: lighting

Manufacturer/Client
Industrial Technology Research
Taiwan

Design
Qisda Creative Design Center
Taiwan
ITRI Green Energy & Environment
Taiwan

Inge Ingwersirup   Corporate Design
Category: print media – corporate design

Manufacturer/Client
Inge Ingwersirup
Germany

Design
Zeichen & Wunder GmbH
Germany

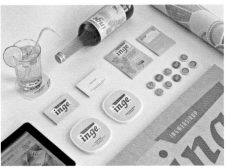

Ingo Maurer   Website
Category: digital media – corporate websites

Manufacturer/Client
Ingo Maurer GmbH
Germany

Design
Heine/Lenz/Zizka Projekte GmbH
Germany

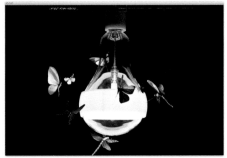

INNO WAVE   Headphone
Category: audio / video

Manufacturer/Client
INNODevice, Inc.
South Korea

Design
INNODesign, Inc.
South Korea

TS K1 System   Server
Category: computer

Manufacturer/Client
Inspur (Beijing) Electronic
China

Design
OREA Technology (Beijing) Co., Ltd.
China

Manufacturer/Client **I**

SS shapewear Shapewear
Category: research+development / professional concepts

Manufacturer/Client
Institute for Information Industry
Taiwan

Design
Taiwan Tech
Taiwan
VALD Design Co., Ltd.
Taiwan

Insuline InsuPad Insulin intake device
Category: medicine / health+care

Manufacturer/Client
Insuline Medical Ltd.
Israel

Design
NewDealDesign LLC
United States of America

EQUAL Chocolate packaging
Category: food

Manufacturer/Client
Interbrand Korea
South Korea

Design
Interbrand Korea
South Korea

SPAA High school
Category: print media – corporate design

Manufacturer/Client
intergram
South Korea

Design
intergram
South Korea

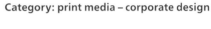

Living in Contrasts Book
Category: print media – product communication

Manufacturer/Client
Interprint GmbH
Germany

Design
LAMOTO
Germany

W Curtain pole
Category: living room / bedroom

Manufacturer/Client
Interstil
Germany

Design
design Frank Greiser
Germany

Manufacturer/Client I

KINETICis5 Bar stool
Category: office / business

Manufacturer/Client
Interstuhl
Büromöbel GmbH & Co. KG
Germany

Design
Phoenix Design
Germany

VINTAGEis5 Conference armchair
Category: office / business

Manufacturer/Client
Interstuhl
Büromöbel GmbH & Co. KG
Germany

Design
Eysing Design Kiel
Germany

Flot LED downlight with backlight
Category: lighting

Manufacturer/Client
Intra lighting
Slovenia

Design
SERGE CORNELISSEN BVBA
Belgium

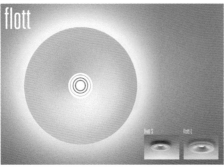

Pipes Professional LED spotlight
Category: lighting

Manufacturer/Client
Intra lighting
Slovenia

Design
SERGE CORNELISSEN BVBA
Belgium

mitu IvyLight Outside luminaire
Category: lighting

Manufacturer/Client
IP44 Schmalhorst GmbH & Co. KG
Germany

Design
Klaus Nolting . ON3D
Germany

IPSA PREMIER LINE Cosmetics packaging
Category: beauty / health / household

Manufacturer/Client
IPSA CO., Ltd.
Japan

Design
Communication Design Laboratory
Japan

Manufacturer/Client **I**

Astell & Kern 120 Hi-Fi Audio Player
Category: audio / video

Manufacturer/Client
IRIVER
South Korea

Design
IRIVER
South Korea

GOLD

JURY STATEMENT

"The reason why we selected the hifi audio player was the amazing attention to details, the particular craftsmanship on the product. It gives a high quality impression. The feeling in relation to the interface and the analog usability, the way you can operate the product, is fantastic as well. It is a product which is very distinguished, well designed and thus a product which deserves an iF gold award."

Astell & Kern 10 Portable DAC
Category: audio / video

Manufacturer/Client
IRIVER
South Korea

Design
IRIVER
South Korea

iRobot Roomba® 800 Robotic vacuum cleaner
Category: household / tableware

Manufacturer/Client
iRobot Corporation
United States of America

Design
iRobot Corporation
United States of America

BAO BAO ISSEY MIYAKE Retail
Category: corporate architecture – shop / showroom

Manufacturer/Client
ISSEY MIYAKE Inc.
Japan

Design
MOMENT
Japan

GOLD

JURY STATEMENT

"We are honouring a showroom by Issey Miyake for the product Bao Bao, which is about a bag concept made of triangular elements. The store interprets this product idea with its simplicity and reduction as well as its geometric specialness. It develops an immense visual energy and long-distance effect. The sculptural quality of the products inside the store is intensified by the background design which was inspired by the same."

BAO BAO ISSEY MIYAKE   Retail
Category: corporate architecture – shop / showroom

Manufacturer/Client
ISSEY MIYAKE Inc.
Japan

Design
MOMENT
Japan

EASY MY BOTTLE   Easy squeeze bottle
Category: packaging formdesign

Manufacturer/Client
IT DESIGN
South Korea

Design
Lee Young Joo
South Korea

TPS   Stairway / platform system
Category: industry / skilled trades

Manufacturer/Client
item Industrietechnik GmbH
Germany

Design
item Industrietechnik GmbH
Germany

Folds of Sea Surface   Installation
Category: corporate architecture – installations in public spaces

Manufacturer/Client
ISSEY MIYAKE Inc.
Japan

Design
Drawing and Manual Inc.
Japan
FabLab Shibuya
Japan

Maschinenleuchten LED   Light fittings
Category: lighting

Manufacturer/Client
item Industrietechnik GmbH
Germany

Design
item Industrietechnik GmbH
Germany

Mi Trainer   Fitness accessory
Category: research+development / professional concepts

Manufacturer/Client
Industrial Technology Research Institute, Taiwan

Design
Lite-on Technology Corp.
Taiwan

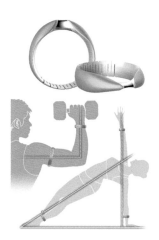

Manufacturer/Client

VSBS Reflective pulse oximeter
Category: research+development / professional concepts

Manufacturer/Client
Industrial Technology Research Institute, Taiwan

Design
Pilotfish
Munich|Amsterdam|Taipei
Taiwan

Linea Basin mixer
Category: bathroom / wellness

Manufacturer/Client
Jaquar & Company Pvt. Ltd.
India

Design
Jaquar & Company Pvt. Ltd.
India

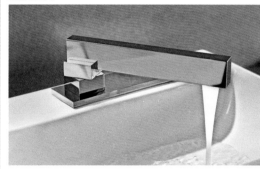

JOY Fitting series
Category: bathroom / wellness

Manufacturer/Client
JADO
Belgium

Design
ARTEFAKT product design
Germany

GOLD

JURY STATEMENT

"This unique mixer tap, characteristically named Joy, convinced us at first glance through its clear lines and proportions as well as the excellent craftsmanship. However, it is the handling and the adjustment of temperature and water volume which turns it into an outstanding product, as with the rising amount of water, the operating cylinders slowly unscrews itself from the fitting."

Boneless Pork Gift box
Category: food

Manufacturer/Client
Jardin De Jade
China

Design
Proad Identity
Taiwan

Rice Cake Rice cake
Category: food

Manufacturer/Client
Jardin De Jade
China

Design
Proad Identity
Taiwan

Manufacturer/Client **J**

JK900 Straight and arc photographic slider
Category: audio / video

Manufacturer/Client
Jawbone Industrial Co., Ltd.
Taiwan

Design
Jawbone Industrial Co., Ltd.
Taiwan

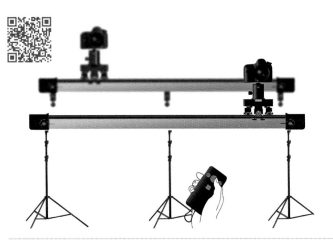

SF- Sensor Fire sensor
Category: buildings

Manufacturer/Client
JBX SYSTEM
Taiwan

Design
GIXIA GROUP Co.
Taiwan

SmartMedicap (7DAY-PRJT1) Medicine bottle with date display
Category: beauty / health / household

Manufacturer/Client
JEILTECH
South Korea

Design
moo:u design
South Korea

GOLD

JURY STATEMENT

"The extraordinary function of this pillbox is particularly delightful: through a certain mechanism the date automatically changes upon opening the box. This is of great use to the customer if ingesting medication on a regular basis is necessary. The box is easy to open and convinces with its appealing graphics. Furthermore, the product does not look like a typical pillbox but rather like an attractive beauty accessory. Also, the well-made shape has a nice heft to it."

Hallasu Glass-bottled drinking water
Category: beverages

Manufacturer/Client
Jeju Special Self-governing
Province Development Corp.
South Korea

Design
Jeju Special Self-governing
Province Development Corp.
South Korea
CROSSPOINT, South Korea
d'ORIGIN, South Korea
JPDC, South Korea

TraffiTower 2.0 Traffic enforcement system
Category: public design

Manufacturer/Client
Jenoptik Robot GmbH
Germany

Design
Jakubowski – Büro für Gestaltung
Germany

Manufacturer/Client **J**

awaseru Marriage ring
Category: leisure / lifestyle

Manufacturer/Client
Jewelry Shalom
Japan

Design
Concertino
Japan

Taste Condiment set
Category: household / tableware

Manufacturer/Client
JIA Inc.
Hong Kong

Design
JIA Inc.
Hong Kong

Solfy F Dental ultrasound scaler
Category: medicine / health+care

Manufacturer/Client
J. Morita Mfg. Corp.
Japan

Design
f/p design GmbH
Germany

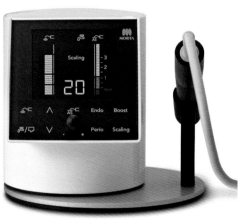

JOBY Mounting device for action cam
Category: leisure / lifestyle

Manufacturer/Client
JOBY Inc.
United States of America

Design
JOBY Inc.
United States of America

Beginning Basin faucet
Category: bathroom / wellness

Manufacturer/Client
JOMOO
Kitchen & Bath Appliances Co., Ltd.
China

Design
JOMOO
Kitchen & Bath Appliances Co., Ltd.
China

Combination pie Storage basin faucet
Category: bathroom / wellness

Manufacturer/Client
JOMOO
Kitchen & Bath Appliances Co., Ltd.
China

Design
JOMOO
Kitchen & Bath Appliances Co., Ltd.
China

Manufacturer/Client **J**

JOURNEY   Basin faucet
Category: bathroom / wellness

Manufacturer/Client
JOMOO
Kitchen & Bath Appliances Co., Ltd.
China

Design
JOMOO
Kitchen & Bath Appliances Co., Ltd.
China

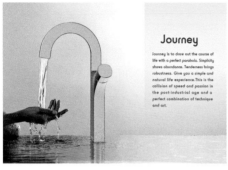

M-shower   Shower system
Category: bathroom / wellness

Manufacturer/Client
JOMOO
Kitchen & Bath Appliances Co., Ltd.
China

Design
SPRGO Design
China

Mus-shower   Shower
Category: bathroom / wellness

Manufacturer/Client
JOMOO
Kitchen & Bath Appliances Co., Ltd.
China

Design
SPRGO Design
China

Magic Sink   Sink
Category: research+development / professional concepts

Manufacturer/Client
JOMOO
Kitchen & Bath Appliances Co., Ltd.
China

Design
JOMOO
Kitchen & Bath Appliances Co., Ltd.
China

TriScale   Kitchen scales
Category: household / tableware

Manufacturer/Client
Joseph Joseph
United Kingdom

Design
Joseph Joseph
United Kingdom

Life-Link   Health monitor system
Category: medicine / health+care

Manufacturer/Client
JoyaTech
China

Design
JoyaTech
China

Manufacturer/Client **J**

FACTOR 21 (27.5) Carbon trail wheels
Category: leisure / lifestyle

Manufacturer/Client
Joy Industrial Co., Ltd.
Taiwan

Design
Joy Industrial Co., Ltd.
Taiwan

i-soft plus Water softener
Category: buildings

Manufacturer/Client
JUDO Wasseraufbereitung GmbH
Germany

Design
KIENLEDESIGN
Germany

AluPocket™ iPhone wall mount
Category: telecommunications

Manufacturer/Client
Just Mobile Ltd.
Taiwan

Design
Tools Design
Denmark

Gum Max Duo™ Universal power pack
Category: telecommunications

Manufacturer/Client
Just Mobile Ltd.
Taiwan

Design
Just Mobile Ltd.
Taiwan

AluPen Twist™ Dual-function stylus
Category: office / business

Manufacturer/Client
Just Mobile Ltd.
Taiwan

Design
Tools Design
Denmark

B-web 96 00 Terminal for time and access control
Category: buildings

Manufacturer/Client
KABA
Germany

Design
INDEED
Germany

Manufacturer/Client **K**

Unit Ceramic furniture
Category: living room / bedroom

Manufacturer/Client
Kähler Design A / S
Denmark

Design
Kähler Design A / S
Denmark

Kamstrup 302 MULTICAL® 302 Heat and cooling meter
Category: buildings

Manufacturer/Client
Kamstrup A / S
Denmark

Design
Kjaerulff Design
Denmark

Kamstrup MULTICAL 302

De Tainan Stijl Book
Category: print media – publishing

Manufacturer/Client
Kan & Lau Design Consultants
Hong Kong

Design
Kan & Lau Design Consultants
Hong Kong

IntertwinedChopstick Tableware
Category: household / tableware

Manufacturer/Client
Kan & Lau Design Consultants
Hong Kong

Design
Kan & Lau Design Consultants
Hong Kong

Sozialbericht Solothurn Image concept
Category: print media – information media

Manufacturer/Client
Kanton Solothurn
Switzerland

Design
André Konrad
Switzerland
Katharina Andes
Switzerland
Martin L. Daester
Switzerland

Window to the East Photography expedition
Category: print media – publishing

Manufacturer/Client
Kaohsiung Museum of Fine Art
Taiwan

Design
ushowdesign Co., Ltd.
Taiwan

Manufacturer/Client **K**

Red Bull Illume   Multisensory packaging
Category: leisure / lifestyle

Manufacturer/Client
Karl Knauer KG, Germany
ROX Asia Consultancy Ltd.
Germany

Design
zooom production GmbH
Austria

KBS Media Facade   Korea's Presidential Election
Category: crossmedia – events

Manufacturer/Client
KBS
South Korea

Design
d'strict
South Korea

NAVIEN 7 Wall pad"   Home network system
Category: buildings

Manufacturer/Client
KD One Co., Ltd.
South Korea

Design
KD Navien Co., Ltd.
South Korea

KMC015 CHEF Diamond White   Kitchen machine
Category: kitchen

Manufacturer/Client
Kenwood Limited
United Kingdom

Design
Kenwood Limited
United Kingdom

LINERO MosaiQ   Backsplash midway system
Category: kitchen

Manufacturer/Client
Kesseböhmer GmbH
Germany

Design
Kesseböhmer GmbH
Germany
GENERATION DESIGN GmbH
Germany

iLook_move   Cosmetic mirror
Category: bathroom / wellness

Manufacturer/Client
KEUCO GmbH & Co. KG
Germany

Design
Tesseraux + Partner
Germany

Manufacturer/Client **K**

meTime_spa Showerboard
Category: bathroom / wellness

Manufacturer/Client
KEUCO GmbH & Co. KG
Germany

Design
Tesseraux + Partner
Germany

Kia Soul Vehicle
Category: transportation design / special vehicles

Manufacturer/Client
Kia Motors Corporation
Germany

Design
Kia Design Team
Germany

KTP420X Tablet press
Category: medicine / health+care

Manufacturer/Client
KILIAN Tableting GmbH
Germany

Design
Budde Industrie Design GmbH
Germany

KfW Stiftung Foundation website
Category: digital media – corporate websites

Manufacturer/Client
KfW Stiftung
Germany

Design
SYZYGY Deutschland GmbH
Germany

Hugdoll Seatbelt positioner
Category: leisure / lifestyle

Manufacturer/Client
KIDU
South Korea

Design
KIDU
South Korea

L595 Detachable handle for cookware
Category: household / tableware

Manufacturer/Client
KIMS HOLDINGS
South Korea

Design
KIMS HOLDINGS
South Korea

Manufacturer/Client **K**

Kinnasand Image brochure
Category: print media – corporate design

Manufacturer/Client
Kinnasand GmbH
Germany

Design
JUNO
Germany

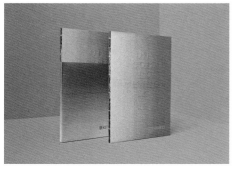

KircherBurkhardt Web Website
Category: digital media – corporate websites

Manufacturer/Client
KircherBurkhardt GmbH
Germany

Design
KircherBurkhardt GmbH
Germany
Digitalwerk GmbH
Austria

Klever S45 Speed electric bicycle
Category: leisure / lifestyle

Manufacturer/Client
Klever Mobility Inc.
Taiwan

Design
MONK Design
Switzerland

KION GROUP Imagebroschüre Image brochure
Category: print media – corporate design

Manufacturer/Client
KION GROUP AG
Germany

Design
3st kommunikation GmbH
Germany
PLATOON KOMMUNIZIERT
Germany

Mobirex MR110Z / MR130Z Mobile impact crusher
Category: transportation design / special vehicles

Manufacturer/Client
Kleemann GmbH
Germany

Design
Dialogform GmbH
Germany

KMS TEAM Company website relaunch
Category: digital media – corporate websites

Manufacturer/Client
KMS TEAM GmbH
Germany

Design
KMS TEAM
Germany

Manufacturer/Client **K**

SAGA-D Disaster helmet
Category: industry / skilled trades

Manufacturer/Client
KMW
South Korea

Design
KMW
South Korea

Knabenkantorei Corporate Design
Category: print media – corporate design

Manufacturer/Client
Knabenkantorei Basel
Switzerland

Design
Schaffner & Conzelmann AG
Switzerland

KOGA BeachRacer Cycle-Crosser
Category: leisure / lifestyle

Manufacturer/Client
Koga B. V.
Netherlands

Design
Koga B. V.
Netherlands

KOGA F3-Serie Urban bike
Category: leisure / lifestyle

Manufacturer/Client
Koga B. V.
Netherlands

Design
Koga B. V.
Netherlands

Moxie Showerhead with speaker
Category: bathroom / wellness

Manufacturer/Client
Kohler China Investment Co., Ltd.
China

Design
Kohler China Investment Co., Ltd.
China

Mira Fluency Tap
Category: bathroom / wellness

Manufacturer/Client
Kohler Mira Ltd.
United Kingdom

Design
Kohler Mira Ltd.
United Kingdom

Manufacturer/Client **K**

Rice brand   Rice packaging
Category: food

Manufacturer/Client
Koike Rice Store
Japan

Design
SAKURA Inc.
Japan
KREO CO., Ltd.
Japan

Tola   Executive conference chair
Category: office / business

Manufacturer/Client
Koleksiyon Mobilya San. A. Ş.
Turkey

Design
f/p design GmbH
Germany

KFA Bericht 2012   Annual Report
Category: print media – publishing

Manufacturer/Client
Kölner Freiwilligen Agentur
Germany

Design
muehlhausmoers
Germany

GEONIC   Surface-design material
Category: material / textiles / wall+floor

Manufacturer/Client
KOLON
South Korea

Design
KOLON
South Korea

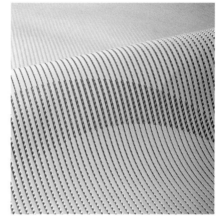

KONE RL20   Elevator car ceiling light
Category: lighting

Manufacturer/Client
KONE Corporation
Finland

Design
KONE Design Solutions
Finland

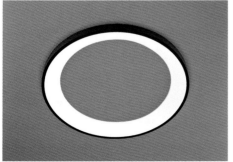

KONE KSP 858   Destination control unit
Category: buildings

Manufacturer/Client
KONE Corporation
Finland

Design
KONE Design Solutions
Finland

Manufacturer/Client **K**

Gazelle Ultimate T3i — Electric bicycle
Category: leisure / lifestyle

Manufacturer/Client
Koninklijke Gazelle N. V.
Netherlands

Design
Koninklijke Gazelle N. V.
Netherlands

skai® cool colors Venezia — Upholstery material
Category: material / textiles / wall+floor

Manufacturer/Client
Konrad Hornschuch AG
Germany

Design
Konrad Hornschuch AG
Germany

Website Konzerthaus — Website
Category: digital media – public service websites

Manufacturer/Client
Konzerthaus Berlin
Germany

Design
m.i.r. media
Germany

Brand&Product Experience 2013 — Design yearbook
Category: print media – corporate communication

Manufacturer/Client
Konrad Hornschuch AG
Germany

Design
Konrad Hornschuch AG
Germany

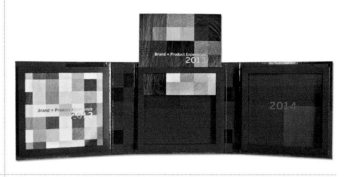

FIRE — Extinguisher holder
Category: office / business

Manufacturer/Client
Konstantin Slawinski
Germany

Design
Formfusion
Germany

SHOOTTER — Website
Category: digital media – microsites

Manufacturer/Client
Kose Corporation
Japan

Design
Hakuhodo
Japan
mount inc.
Japan

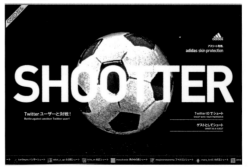

Manufacturer/Client **K**

PET lite 9.9 carbonated Lightweight bottle
Category: beverages

Manufacturer/Client
Krones AG
Germany

Design
Krones AG
Germany

Krones MLA Mobile line assistant
Category: product interfaces

Manufacturer/Client
Krones AG
Germany

Design
Krones AG
Germany

olleh tv smart TV interface
Category: product interfaces

Manufacturer/Client
kt media hub
South Korea

Design
kt media hub
South Korea

DUAL GRATER Dual grater
Category: household / tableware

Manufacturer/Client
Kuhn Rikon AG
Switzerland

Design
Kuhn Rikon AG
Switzerland

DUAL KNIFE SHARPENER Dual knife sharpener
Category: household / tableware

Manufacturer/Client
Kuhn Rikon AG
Switzerland

Design
Kuhn Rikon AG
Switzerland

LBR iiwa Lightweight robot
Category: industry / skilled trades

Manufacturer/Client
KUKA AG
Germany

Design
Selic Industriedesign
Germany
KUKA AG
Germany

Manufacturer/Client **K**

ECSTA PS91   Tire
Category: transportation design / special vehicles

Manufacturer/Client
Kumho Tire Co.
South Korea

Design
Kumho Tire Co.
South Korea

SOLUS HS51   Tire
Category: transportation design / special vehicles

Manufacturer/Client
Kumho Tire Co.
South Korea

Design
Kumho Tire Co.
South Korea

Ein Arbeitsbuch   Annual Report 2012
Category: print media – corporate communication

Manufacturer/Client
Kuoni Reisen Holding AG
Switzerland

Design
büroecco
Kommunikationsdesign GmbH
Germany

GOLD

JURY STATEMENT

"We have seen a lot of annual reports and they are not known for their creativity. This report is a great example of how important design is to business. Design fundamentals were strictly used to improve the annual report, to make it a good size, to make it lightweight, to make it navigational, to make it easy to flip through. The typography, the great grid system, the creative info graphics elevate this annual report to a prototype."

Poemstory   Book
Category: print media – publishing

Manufacturer/Client
Kuoyenting
Taiwan

Design
Kuoyenting
Taiwan

Kverneland Vicon   EXPERT 432F mower
Category: transportation design / special vehicles

Manufacturer/Client
Kverneland Group
Denmark

Design
VanBerlo B. V.
Netherlands

Manufacturer/Client **K**

Mirka DEROS Electric sander
Category: industry / skilled trades

Manufacturer/Client
KWH Mirka Ltd.
Finland

Design
Veryday
Sweden

Ring Pointer Mini presenter
Category: computer

Manufacturer/Client
KYE SYSTEMS Corp. (Genius)
Taiwan

Design
KYE SYSTEMS Corp.
Taiwan

DIGNO M Smartphone
Category: telecommunications

Manufacturer/Client
KYOCERA Corporation
Japan

Design
KYOCERA Corporation
Japan

Wells Series1 Water purifier
Category: kitchen

Manufacturer/Client
kyowon L & C
South Korea

Design
kyowon L & C
South Korea

PURECE Skin massage unit
Category: bathroom / wellness

Manufacturer/Client
kyowon L & C
South Korea

Design
kyowon L & C
South Korea

Cable Case for iPhone 5S iPhone accessory
Category: telecommunications

Manufacturer/Client
LAB.C
South Korea

Design
PlusX
South Korea

Manufacturer/Client **L**

LACO ABSOLUTE QUARZ Wrist watch
Category: leisure / lifestyle

Manufacturer/Client
LACO Uhrenmanufaktur GmbH
Germany

Design
LACO Uhrenmanufaktur GmbH
Germany

Faszinarium Stör Exhibition
Category: corporate architecture – exhibition / trade fair

Manufacturer/Client
Landesanglerverband
Brandenburg e.V.
Germany

Design
stories within architecture
Germany

Gmund For Food Chocolate
Category: food

Manufacturer/Client
Landor and Associates
Germany

Design
Landor and Associates
Germany

Langenscheidt Corporate Design
Category: print media – corporate design

Manufacturer/Client
Langenscheidt GmbH & Co. KG
Germany

Design
KW43 BRANDDESIGN
Germany

WinFast WS1000 Graphics workstation
Category: computer

Manufacturer/Client
Leadtek Research Inc.
Taiwan

Design
Leadtek Research Inc.
Taiwan

XOOMINAIRE Modular linear LED luminaire
Category: lighting

Manufacturer/Client
LED Linear GmbH
Germany

Design
LED Linear GmbH
Germany

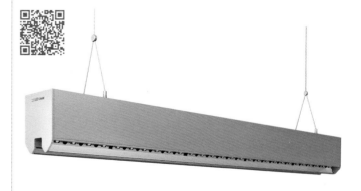

Manufacturer/Client **L**

Xtend+ Illuminated top shelf
Category: **kitchen**

Manufacturer/Client
LEICHT Küchen AG
Germany

Design
LEICHT Küchen AG
Germany

VEGA 12 Trailed sprayer
Category: **transportation design / special vehicles**

Manufacturer/Client
LEMKEN GmbH & Co. KG
Germany

Design
Budde Industrie Design GmbH
Germany

ThinkVision LT2934z 29" panoramic monitor
Category: **computer**

Manufacturer/Client
Lenovo
United States of America

Design
Lenovo
United States of America

63342 Solar Fit Solar personal analysis scale
Category: **bathroom / wellness**

Manufacturer/Client
Leifheit AG
Germany

Design
Leifheit AG
Germany

ThinkCentre E93z All-in-one computer
Category: **computer**

Manufacturer/Client
Lenovo
United States of America

Design
Lenovo
United States of America

GOLD

JURY STATEMENT

"We chose this monitor design particularly because of the successful use of a landscape aspect ratio, creating one dramatic single screen. We liked the simplicity of the camera security and we see this as being the best example of the evolution of the LENOVO Think design language. It is not a series as some of the other examples. It is a little more evolved and playful and overall the proportions and details are very, very good."

Manufacturer/Client **L**

ThinkPad 8 & Cover Tablet
Category: computer

Manufacturer/Client
Lenovo
United States of America

Design
Lenovo
United States of America

ThinkPad X1 Carbon Ultrabook
Category: computer

Manufacturer/Client
Lenovo
United States of America

Design
Lenovo
United States of America

Aura User Interface
Category: product interfaces

Manufacturer/Client
Lenovo (Beijing) Ltd.
China

Design
Lenovo (Beijing) Ltd.
China

K900 Smartphone
Category: telecommunications

Manufacturer/Client
Lenovo (Beijing) Ltd.
China

Design
Lenovo (Beijing) Ltd.
China

Horizon table PC Table PC
Category: computer

Manufacturer/Client
Lenovo (Beijing) Ltd.
China

Design
Lenovo (Beijing) Ltd.
China

YOGA TABLET 8 Tablet
Category: computer

Manufacturer/Client
Lenovo (Beijing) Ltd.
China

Design
Lenovo (Beijing) Ltd.
China

Manufacturer/Client **L**

SIMBA DVD case
Category: print media – product communication

Manufacturer/Client
Lernwelt Wolfpassing
Austria

Design
Stefan Radinger
Austria

LYA KLENZER TRAVELERS BAG Cleanser packaging
Category: beauty / health / household

Manufacturer/Client
leyess
South Korea

Design
leyess
South Korea

G Flex Smartphone
Category: telecommunications

Manufacturer/Client
LG Electronics Inc.
South Korea

Design
LG Electronics Inc.
South Korea

Baum zu glauben Christmas book
Category: print media – publishing

Manufacturer/Client
Lesmo GmbH & Co. KG
Germany

Design
Lesmo GmbH & Co. KG
Germany

Leica S-Magazin 3 Magazine
Category: print media – publishing

Manufacturer/Client
LFI Photographie GmbH
Germany

Design
Tom Leifer Design
Germany
LFI Photographie GmbH
Germany

GOLD

JURY STATEMENT

"The new approach to the interface on this smartphone is great. Putting the control on the back and adopting a curved display add immensely to the experience of using this phone. This is an excellent product in terms of its ergonomics and quality."

Manufacturer/Client **L**

CM2540 Awake micro audio system
Category: audio / video

Manufacturer/Client
LG Electronics Inc.
South Korea

Design
LG Electronics Inc.
South Korea
V2 Studios Ltd.
United Kingdom

LB8700 LED TV
Category: audio / video

Manufacturer/Client
LG Electronics Inc.
South Korea

Design
LG Electronics Inc.
South Korea

NP8540 Wireless multi-room speaker
Category: audio / video

Manufacturer/Client
LG Electronics Inc.
South Korea

Design
LG Electronics Inc.
South Korea

LA9700 Ultra HD TV
Category: audio / video

Manufacturer/Client
LG Electronics Inc.
South Korea

Design
LG Electronics Inc.
South Korea

NB5530 Super slim soundbar
Category: audio / video

Manufacturer/Client
LG Electronics Inc.
South Korea

Design
LG Electronics Inc.
South Korea

PG65 Smart portable LED projector
Category: audio / video

Manufacturer/Client
LG Electronics Inc.
South Korea

Design
LG Electronics Inc.
South Korea

Manufacturer/Client **L**

SoundPlate™ (LAP340) 4.1-ch speaker system
Category: audio / video

Manufacturer/Client
LG Electronics Inc.
South Korea

Design
LG Electronics Inc.
South Korea

LG G2 Smartphone
Category: telecommunications

Manufacturer/Client
LG Electronics Inc.
South Korea

Design
LG Electronics Inc.
South Korea

Optimus it (L-05E) Smartphone
Category: telecommunications

Manufacturer/Client
LG Electronics Inc.
South Korea

Design
LG Electronics Inc.
South Korea

Pocket Photo (PD239) Portable mobile printer
Category: telecommunications

Manufacturer/Client
LG Electronics Inc.
South Korea

Design
LG Electronics Inc.
South Korea

VU: 3 (LG-F300) Smartphone
Category: telecommunications

Manufacturer/Client
LG Electronics Inc.
South Korea

Design
LG Electronics Inc.
South Korea

WCP300 Wireless mini charging pad
Category: telecommunications

Manufacturer/Client
LG Electronics Inc.
South Korea

Design
LG Electronics Inc.
South Korea

Manufacturer/Client **L**

34UM95   34" 21:9 LCD Monitor
Category: **computer**

Manufacturer/Client
LG Electronics Inc.
South Korea

Design
LG Electronics Inc.
South Korea

L04D   Mini Wi-Fi router
Category: **computer**

Manufacturer/Client
LG Electronics Inc.
South Korea

Design
LG Electronics Inc.
South Korea

Z940   Ultrabook
Category: **computer**

Manufacturer/Client
LG Electronics Inc.
South Korea

Design
LG Electronics Inc.
South Korea

LG GS-100 Headset Package   Accessory packaging
Category: **consumer electronics**

Manufacturer/Client
LG Electronics Inc.
South Korea

Design
LG Electronics Inc.
South Korea

ARTEMIS HP (MW233SK)   Microwave oven
Category: **kitchen**

Manufacturer/Client
LG Electronics Inc.
South Korea

Design
LG Electronics Inc.
South Korea

ARTEMIS VP (MW233RW)   Microwave oven
Category: **kitchen**

Manufacturer/Client
LG Electronics Inc.
South Korea

Design
LG Electronics Inc.
South Korea

Manufacturer/Client **L**

P Kaiser(GSP545NSYZ) Side-by-side Refrigerator
Category: kitchen

Manufacturer/Client
LG Electronics Inc.
South Korea

Design
LG Electronics Inc.
South Korea

Airwasher (LAW-A048AS) Humidifier
Category: household / tableware

Manufacturer/Client
LG Electronics Inc.
South Korea

Design
LG Electronics Inc.
South Korea

Steamer (H-63HSW) Humidifier
Category: household / tableware

Manufacturer/Client
LG Electronics Inc.
South Korea

Design
LG Electronics Inc.
South Korea

ARTCOOL Slim Air conditioner
Category: buildings

Manufacturer/Client
LG Electronics Inc.
South Korea

Design
LG Electronics Inc.
South Korea

ARTCOOL Stylist Air conditioner
Category: buildings

Manufacturer/Client
LG Electronics Inc.
South Korea

Design
LG Electronics Inc.
South Korea

Rio Convertible air conditioner
Category: buildings

Manufacturer/Client
LG Electronics Inc.
South Korea

Design
LG Electronics Inc.
South Korea

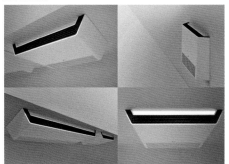

Manufacturer/Client **L**

New Stainless Steel Coating for refrigerator
Category: material / textiles / wall+floor

Manufacturer/Client
LG Electronics Inc.
South Korea

Design
LG Electronics Inc.
South Korea

CLIP100 Router
Category: computer

Manufacturer/Client
LG innotek
South Korea

Design
LG innotek
South Korea

Licon Liflex II 444 Highspeed machining center
Category: industry / skilled trades

Manufacturer/Client
Licon mt GmbH & Co. KG
Germany

Design
defortec GmbH
Germany

Dual Motion Window Dual window device
Category: buildings

Manufacturer/Client
LG Hausys
South Korea

Design
LG Hausys
South Korea

Libratone Loop Audio speaker
Category: audio / video

Manufacturer/Client
Libratone A/S
Denmark

Design
Libratone A/S
Denmark

LH60 Litronic Material handler
Category: transportation design / special vehicles

Manufacturer/Client
Liebherr
Germany

Design
Liebherr
Germany

Manufacturer/Client **L**

LIFEHAMMER EVOLUTION   Safety hammer
Category: transportation design / special vehicles

Manufacturer/Client
Life Safety Products BV
Netherlands

Design
Spark Design & Innovation
Netherlands

Power Charger   Portable power charger
Category: telecommunications

Manufacturer/Client
Lifetrons Switzerland AG
Switzerland

Design
Lifetrons Switzerland AG
Switzerland

WiFi Router / Charger   Portable router and charger
Category: computer

Manufacturer/Client
Lifetrons Switzerland AG
Switzerland

Design
Lifetrons Switzerland AG
Switzerland

Luggage Scale   Rechargeable luggage scale
Category: office / business

Manufacturer/Client
Lifetrons Switzerland AG
Switzerland

Design
Lifetrons Switzerland AG
Switzerland

CARBON BLACK LINE   Linear LED lighting
Category: lighting

Manufacturer/Client
LIGHTperMETER
Belgium

Design
LIGHTperMETER
Belgium

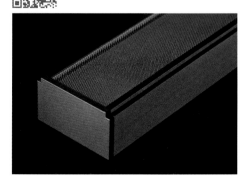

KeyKeg 20 Slimline   One-way keg
Category: beverages

Manufacturer/Client
Lightweight Containers
Netherlands

Design
Lightweight Containers
Netherlands

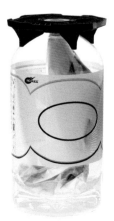

Manufacturer/Client **L**

Linde Fascinating Gases iPad App
Category: digital media – mobile applications

Manufacturer/Client
Linde AG
Germany

Design
Peter Schmidt Group
Germany
amp – audible brand
and corporate communication
Germany

Linde R16 Reach truck
Category: transportation design / special vehicles

Manufacturer/Client
Linde Material Handling GmbH
Germany

Design
Linde Material Handling GmbH
Germany
Porsche Engineering Group GmbH
Germany

Green in Black Walking shoes
Category: leisure / lifestyle

Manufacturer/Client
LI-NING (China) Sports Goods Co.,
China

Design
LI-NING (China) Sports Goods Co.,
China

Entschlossenheit Annual Report 2012
Category: print media – corporate communication

Manufacturer/Client
Linde AG
Germany

Design
hw.design GmbH
Germany

LINE ID CARD CASE ID card and case
Category: office / business

Manufacturer/Client
LINE
South Korea

Design
LINE
South Korea

Manfrotto MTPIXI-B Mini tripod
Category: audio / video

Manufacturer/Client
Lino Manfrotto + Co. S. p. A.
Italy

Design
Cesare Carlesso – Designer
Italy

Manufacturer/Client **L**

Slow and Flow   Tea party
Category: crossmedia – events

Manufacturer/Client
Lin's Ceramics Studio
Taiwan

Design
Studio Qiao
Taiwan

A Happy Excursion   Packaging design
Category: leisure / lifestyle

Manufacturer/Client
LIULIGONGFANG
China

Design
LIULIGONGFANG
China

LMS SoundBrush   Tool for sound visualization
Category: industry / skilled trades

Manufacturer/Client
LMS Instruments B. V.
Netherlands

Design
npk design
Netherlands

LAO SHE TEA SHOP   Tea packaging
Category: beverages

Manufacturer/Client
linshaobin design
China

Design
linshaobin design
China

Kanzlei LM   Corporate brand
Category: crossmedia – corporate design

Manufacturer/Client
LM Audit und Tax GmbH
Germany

Design
Morgen Digital
Germany

R1B2   Cordless riveter
Category: industry / skilled trades

Manufacturer/Client
LOBTEX Co., Ltd.
Japan

Design
LOBTEX Co., Ltd.
Japan

Manufacturer/Client **L**

3D Orchestra Speaker — Sound system
Category: audio / video

Manufacturer/Client
Loewe AG
Germany

Design
Design 3
Germany

TOUCHdown — Table handswitch
Category: office / business

Manufacturer/Client
LOGICDATA GmbH
Austria

Design
LOGICDATA GmbH
Austria

Logitech Broadcaster — Travel case
Category: consumer electronics

Manufacturer/Client
Logitech
United States of America

Design
Daylight Design Inc.
United States of America

GOLD

JURY STATEMENT

"This contribution is a great example for being very inventive in that after purchase the packaging supports the use of the product itself. This case is very innovative because it supports the usability of the product – the camera. This level of multifunctionality, paired with how the packaging comes together with the product, is exemplary and clever."

Logitech Broadcaster — Wi-Fi webcam for HD video
Category: audio / video

Manufacturer/Client
Logitech
United States of America

Design
Daylight Design Inc.
United States of America

Logitech Speakers Z600 — Bluetooth speaker
Category: audio / video

Manufacturer/Client
Logitech
United States of America

Design
NONOBJECT
United States of America

Manufacturer/Client **L**

FabricSkin Folio iPad keyboard folio
Category: computer

Manufacturer/Client
Logitech Europe
Switzerland

Design
Design Partners
Ireland

Logitech T650 Touchpad Touchpad
Category: computer

Manufacturer/Client
Logitech Europe
Switzerland

Design
Design Partners
Ireland

Logitech G602 Gaming mouse
Category: computer

Manufacturer/Client
Logitech Gateway
United States of America

Design
Design Partners
Ireland

Logitech Keyboard K810 Keyboard
Category: computer

Manufacturer/Client
Logitech Europe
Switzerland

Design
Design Partners
Ireland

Ultrathin TouchMouse T630 Mouse
Category: computer

Manufacturer/Client
Logitech Europe
Switzerland

Design
Design Partners
Ireland

Bike Light Bike light / smart beam
Category: leisure / lifestyle

Manufacturer/Client
Lord Benex International Co., Ltd.
Taiwan

Design
Lord Benex International Co., Ltd.
Taiwan

Manufacturer/Client **L**

LOTTE GUM Gum set packaging
Category: food

Manufacturer/Client
LOTTE Confectionery
South Korea

Design
cdupartners
South Korea
LOTTE Confectionery
South Korea

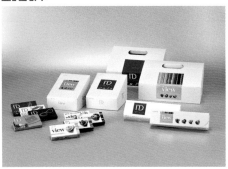

LOTTE GUM Gum set packaging
Category: packaging formdesign

Manufacturer/Client
LOTTE Confectionery
South Korea

Design
cdupartners
South Korea
LOTTE Confectionery
South Korea

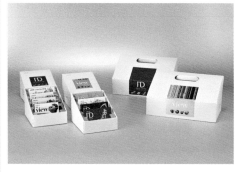

LP Circle Fixture series
Category: lighting

Manufacturer/Client
Louis Poulsen Lighting A / S
Denmark

Design
Louis Poulsen Lighting A / S
Denmark

MicroLine 2000 PCB laser cutting system
Category: industry / skilled trades

Manufacturer/Client
LPKF AG
Germany

Design
stephan gahlow produktgestaltung
Germany

Velvet Decanter Mouthblown glass decanter
Category: household / tableware

Manufacturer/Client
LSA International
United Kingdom

Design
LSA International
United Kingdom

fool Table lamp
Category: lighting

Manufacturer/Client
lumini
Brazil

Design
lumini
Brazil

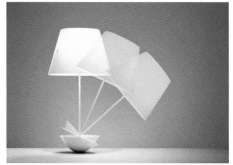

Manufacturer/Client **L**

vinte2 Table and pendant lamps
Category: lighting

Manufacturer/Client
lumini
Brazil

Design
lumini
Brazil

SF Prep Earthquake kit service
Category: research+development / professional concepts

Manufacturer/Client
LUNAR
United States of America

Design
LUNAR
United States of America

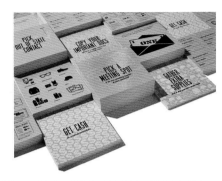

NOVA Climbing wall
Category: research+development / professional concepts

Manufacturer/Client
LUNAR
Germany

Design
LUNAR
Germany

GOLD

JURY STATEMENT

"A climbing wall which can be integrated into the interior design of living spaces is something completely new and an excellent design achievement. Through the integrated light effect at the back this product has a unique modern feel. It is not an imitation of nature for an artificial space; rather it offers the same range of functions as a natural climbing wall, which makes it truly unique."

VELA Stationary bike
Category: research+development / professional concepts

Manufacturer/Client
LUNAR
Germany

Design
LUNAR
Germany

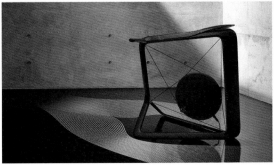

SimiLED ART LED light
Category: lighting

Manufacturer/Client
Luxuni GmbH
Germany

Design
Phenix lighting (Xiamen) Co., Ltd.
China

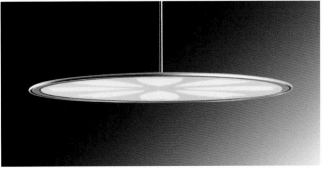

Manufacturer/Client **L**

LYA BALL KLENZER   Cleanser packaging
Category: beauty / health / household

Manufacturer/Client	Design
LYA NATURE	leyess
South Korea	South Korea

GOLD

JURY STATEMENT

"The design of this cosmetic packaging is very minimal. The beautiful, pure design is especially impressive through its pure shape. As soon as you touch it, the texture offers a very subtle and nice feeling that creates an excellent haptic experience for the user while still presenting an appropriate packaging for the product."

Lytro   Light field camera
Category: audio / video

Manufacturer/Client	Design
Lytro, Inc.	Lytro, Inc.
United States of America	United States of America

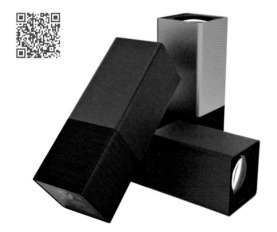

BLUNT™ Golf_G1   Umbrella
Category: leisure / lifestyle

Manufacturer/Client	Design
Madeblunt Ltd.	Madeblunt Ltd.
New Zealand	New Zealand

Combine Bag   Bag
Category: leisure / lifestyle

Manufacturer/Client	Design
Made in China by Boldº	Boldº_a design company
Brazil	Brazil

Magic Bubble   Logo and Visual Identity
Category: print media – corporate design

Manufacturer/Client	Design
Magic Bubble	Dongdao Design Co., Ltd.
China	China

Manufacturer/Client **M**

Makita Power Tools   Every day the right tools.
Category: print media – product communication

Manufacturer/Client
Makita Werkzeug GmbH
Germany

Design
Leo Burnett GmbH
Germany

nexOLED   Pendant
Category: lighting

Manufacturer/Client
Maltani lighting
South Korea

Design
Maltani lighting
South Korea
Seoul Women's Univ.
South Korea

X-road   LED pole
Category: lighting

Manufacturer/Client
Maltani lighting
South Korea

Design
Maltani lighting
South Korea
Seoul Women's Univ.
South Korea

Mammut   Packaging design
Category: leisure / lifestyle

Manufacturer/Client
Mammut Sports Group AG
Switzerland

Design
Hauser Lacour
Germany
MindDesign
United Kingdom

Viale BRT   City bus
Category: transportation design / special vehicles

Manufacturer/Client
Marcopolo S/A
Brazil

Design
DESIGN CENTER MARCOPOLO
Brazil

Onno Jetel (KC 41)   Motorboat
Category: transportation design / special vehicles

Manufacturer/Client
Marina Brodersby
Germany

Design
Onno Jetel e. K.
Germany

Manufacturer/Client **M**

CIRCULAR POL XXL Hanging lamp
Category: lighting

Manufacturer/Client
MARTINELLI LUCE S. p. A.
Italy

Design
MARTINELLI LUCE S. p. A.
Italy

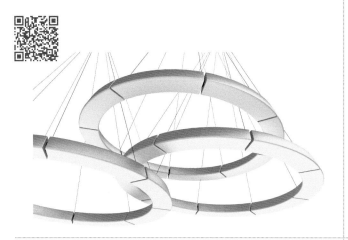

Qube² Pocket speaker
Category: audio / video

Manufacturer/Client
Matrix Audio Ltd.
Canada

Design
Matrix Audio Ltd.
Canada

SB-WS52-SHLW Bluetooth 4.0 Stereo Headset
Category: telecommunications

Manufacturer/Client
MAVIN TECHNOLOGY Inc.
Taiwan

Design
SOFTBANK BB Corp.
Japan

FLUTE Universal transmitter
Category: buildings

Manufacturer/Client
MASTER S. p. A.
Italy

Design
Mario Mazzer architect | designer
Italy

CROSSMAX Shoe
Category: leisure / lifestyle

Manufacturer/Client
MAVIC S. A. S.
France

Design
MAVIC S. A. S.
France

MAXU Stairs Special stairs
Category: buildings

Manufacturer/Client
MAXU Technology Inc.
China

Design
MAXU
Technology European Office
France

Manufacturer/Client **M**

Easy Morning Campaign
Category: print media – advertising media

Manufacturer/Client
McDonald's
Austria

Design
DDB Tribal Wien
Austria

GOLD

JURY STATEMENT

"The key strength of this ad campaign is the clever and witty approach to show the difficulties in starting the day. McDonald's key message, how easy life can be with the convenience of their food in the morning, comes across really well although no food is shown and no remark on taste or quality is given. These very creative and stunning installations with items of clothing are so inviting that you would love to visit the gallery."

McMission Augmented Reality App
Category: digital media – mobile applications

Manufacturer/Client
McDonald's Deutschland Inc.
Germany

Design
LessingvonKlenze
Germany
Heye GmbH
Germany

McDonald's CR-Report 2012 Corporate Responsibility Report
Category: print media – corporate communication

Manufacturer/Client
McDonald's Deutschland Inc.
Germany

Design
LessingvonKlenze
Germany
Heye GmbH
Germany

MEDION® LIFE® MD83867 Wi-Fi speaker
Category: audio / video

Manufacturer/Client
Medion AG
Germany

Design
Medion AG
Germany

MEDION® LIFETAB® E7312 7" tablet
Category: computer

Manufacturer/Client
Medion AG
Germany

Design
Medion Designteam
Germany

Manufacturer/Client **M**

BSC 9.1 S Surgical tool
Category: medicine / health+care

Manufacturer/Client
Medtronic
United States of America

Design
Continuum
United States of America

Die Melitta® Cafina® Super automatic coffee machine
Category: kitchen

Manufacturer/Client
Melitta
SystemService GmbH & Co. KG
Germany

Design
Indeed Innovation GmbH
Germany

www.menlosystems.com Website
Category: digital media – corporate websites

Manufacturer/Client
Menlo Systems GmbH
Germany

Design
Zum Kuckuck GmbH & Co. KG
Germany

MENNEKES Wallbox Wall box for EV charging
Category: transportation design / special vehicles

Manufacturer/Client
MENNEKES
Elektrotechnik GmbH & Co. KG
Germany

Design
Kiska GmbH
Austria

Mercury Marine ERC Remote control
Category: transportation design / special vehicles

Manufacturer/Client
Mercury Marine
United States of America

Design
BMW Group DesignworksUSA
United States of America

Ohi Mobile healthcare device
Category: research+development / professional concepts

Manufacturer/Client
Merit Co., Ltd.
South Korea

Design
Merit Co., Ltd.
South Korea

Manufacturer/Client **M**

Polaris Bluetooth headset
Category: audio / video

Manufacturer/Client
Merry Electronics Co., Ltd.
Taiwan

Design
Merry Electronics Co., Ltd.
Taiwan

Gemma Personal sound amplification
Category: medicine / health+care

Manufacturer/Client
Merry Electronics Co., Ltd.
Taiwan

Design
Merry Electronics Co., Ltd.
Taiwan

M-Pure Frame design
Category: buildings

Manufacturer/Client
Merten GmbH
Germany

Design
eliumstudio
France

DefiMonitor EVO Defibrillator / monitor
Category: medicine / health+care

Manufacturer/Client
Metrax GmbH
Germany

Design
Metrax GmbH
Germany

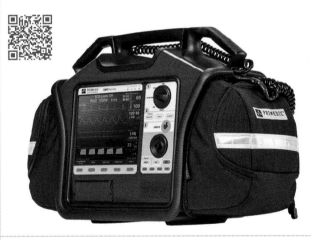

METRO Handelslexikon Retail compendium
Category: print media – information media

Manufacturer/Client
METRO AG
Germany

Design
grintsch communications
GmbH & Co. KG
Germany

Metso MR Moisture Biomass moisture analyzer
Category: industry / skilled trades

Manufacturer/Client
Metso Corporation
Finland

Design
Metso Corporation
Finland

205

Manufacturer/Client **M**

EtaCutII/EtaShredZZ Scrap shear and shredder
Category: industry / skilled trades

Manufacturer/Client
Metso Lindemann
Germany

Design
i/i/d
Institut für Integriertes Design
Germany

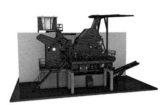

Metso Lokotrack LT106 Mobile crushing plant
Category: transportation design / special vehicles

Manufacturer/Client
Metso Minerals, Inc.
Finland

Design
Metso Minerals, Inc.
Finland

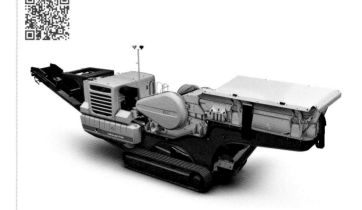

Nachhaltigkeitsbroschüre Brochure
Category: print media – corporate communication

Manufacturer/Client
Meyer & Meyer
Holding GmbH & Co. KG
Germany

Design
red pepper
Germany

COLOR RECOVER™ Hair color coating device
Category: bathroom / wellness

Manufacturer/Client
Michel Mercier
Israel

Design
Aran Research & Development
Israel

Xbox One Chat Headset Gaming headset
Category: computer

Manufacturer/Client
Microsoft
United States of America

Design
Microsoft
United States of America

Xbox One Game Controller Game controller
Category: computer

Manufacturer/Client
Microsoft
United States of America

Design
Microsoft
United States of America

Manufacturer/Client **M**

PG 8055-PG 8060 Baureihe   Professional dishwasher
Category: kitchen

Manufacturer/Client
Miele & Cie. KG
Germany

Design
Miele & Cie. KG
Germany

Ideenfunken-Camp   Cross media event
Category: crossmedia – events

Manufacturer/Client
Milla & Partner
Germany

Design
Milla & Partner Innovationslabor
Germany

miscea LIGHT®   Sensor faucet and dispenser system
Category: bathroom / wellness

Manufacturer/Client
miscea GmbH
Germany

Design
miscea GmbH
Germany

2LINES   Eyeglass frame
Category: leisure / lifestyle

Manufacturer/Client
MiGlaz Design Co., Ltd.
Thailand

Design
MiGlaz Design Co., Ltd.
Thailand

MINOX BL 8 x 33 HD   Binoculars
Category: leisure / lifestyle

Manufacturer/Client
MINOX GmbH
Germany

Design
Volkswagen Design
Germany

Misfit Shine   Personal activity tracker
Category: medicine / health+care

Manufacturer/Client
Misfit Wearables Corporation
United States of America

Design
Misfit Wearables Corporation
United States of America
Pearl Studios Inc.
Canada

Manufacturer/Client **M**

Mio MiVue M350   Motor driving recorder
Category: transportation design / special vehicles

Manufacturer/Client
MiTAC International Corp.
Taiwan

Design
MiTAC International Corp.
Taiwan

Cyclo500   GPS bike device
Category: leisure / lifestyle

Manufacturer/Client
MiTAC International Corp.
Taiwan

Design
MiTAC International Corp.
Taiwan

Lossnay VL   Ventilation fan
Category: buildings

Manufacturer/Client
Mitsubishi Electric Corporation
Japan

Design
Mitsubishi Electric Corporation
Japan

Mivue 568   Driving recorder
Category: transportation design / special vehicles

Manufacturer/Client
MiTAC International Corp.
Taiwan

Design
MiTAC International Corp.
Taiwan

L130   Tough tablet
Category: computer

Manufacturer/Client
MiTAC International Corp.
Taiwan

Design
MiTAC International Corp.
Taiwan

MITSUBISHI GOT 2000Series   Information terminal
Category: industry / skilled trades

Manufacturer/Client
Mitsubishi Electric Corporation
Japan

Design
Mitsubishi Electric Corporation
Japan

Manufacturer/Client **M**

Mitsubishi Electric   Magazine
Category: print media – corporate communication

Manufacturer/Client
Mitsubishi Electric Europe B. V.
Germany

Design
KW43 BRANDDESIGN
Germany

Mobic   Cubicle devices
Category: bathroom / wellness

Manufacturer/Client
MOBIC
South Korea

Design
DESIGN S
South Korea

CompactMask   Respirator
Category: industry / skilled trades

Manufacturer/Client
Moldex-Metric AG & Co. KG
Germany

Design
Moldex-Metric AG & Co. KG
Germany

uni Promark VIEW   Highlighter
Category: office / business

Manufacturer/Client
MITSUBISHI PENCIL Co., Ltd.
Japan

Design
MITSUBISHI PENCIL Co., Ltd.
Japan

NOMADO   Mobile EDV station
Category: office / business

Manufacturer/Client
MOBICA+
Germany

Design
Design Ballendat Germany
Germany

Pétlas   Wall covering
Category: material / textiles / wall+floor

Manufacturer/Client
Mosarte
Brazil

Design
Mosarte
Brazil

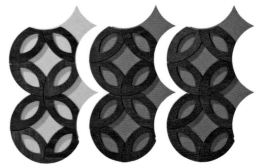

Manufacturer/Client **M**

TLKR Two-Way Radio Series Consumer two-way radios
Category: telecommunications

Manufacturer/Client
Motorola Solutions
United States of America

Design
Motorola Solutions
United States of America

MC40 Mobile Computer Handheld mobile computer
Category: industry / skilled trades

Manufacturer/Client
Motorola Solutions
United States of America

Design
Motorola Solutions
United States of America

MP6000 Scanner/Scale Multi-plane bioptic imager
Category: industry / skilled trades

Manufacturer/Client
Motorola Solutions
United States of America

Design
Motorola Solutions
United States of America

Pomos Fruit spoon
Category: household / tableware

Manufacturer/Client
MTT Integration Corp.
Anguilla

Design
Nova Design (Shanghai) Ltd.
China

Horst und Edeltraut Magazine
Category: print media – publishing

Manufacturer/Client
muehlhausmoers
Germany

Design
muehlhausmoers
Germany

MuKK Store guidance system
Category: corporate architecture – communication media in architecture and public spaces

Manufacturer/Client
MuKK GmbH
Germany

Design
cyclos-design GmbH
Germany

Manufacturer/Client **M**

MGK Siegen Corporate Design
Category: print media – corporate design

Manufacturer/Client
Museum für Gegenwartskunst
Siegen
Germany

Design
Hauser Lacour
Germany

Sprinkle Drinking Water Drinking water bottle
Category: beverages

Manufacturer/Client
M. Water Co., Ltd.
Thailand

Design
Cerebrum Design Co., Ltd.
Thailand

SLIDE Shoe and file cabinet
Category: living room / bedroom

Manufacturer/Client
NAOS Design
Germany

Design
archipool.de
Germany

Stuckvilla Villa Stuck anniversary magazin
Category: print media – publishing

Manufacturer/Client
Museum Villa Stuck
Germany

Design
KMS TEAM
Germany

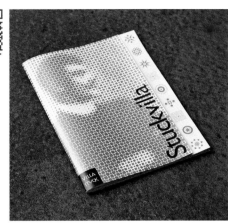

Manhole self-help system Self-help system
Category: research+development / professional concepts

Manufacturer/Client
Nanjing HuaQi
Visual Design Co., Ltd.
China

Design
Nanjing HuaQi
Visual Design Co., Ltd.
China

MMCA Corporate Identity
Category: print media – corporate design

Manufacturer/Client
National Museum of Contemporary Art, Korea
South Korea

Design
INFINITE
South Korea

211

Manufacturer/Client **N**

Viking   Multimedia installation
Category: corporate architecture – installations in public spaces

Manufacturer/Client
National Museum of Denmark
Denmark

Design
ATELIER BRÜCKNER
Germany
Shilo
United States of America

Hangeul Campaign Website   Microsite
Category: digital media – microsites

Manufacturer/Client
NAVER Corp.
South Korea

Design
NAVER Corp.
South Korea

Data Center 'Gak' BI   Brand identity
Category: print media – corporate design

Manufacturer/Client
NAVER Corp.
South Korea

Design
NAVER Corp.
South Korea

Data Center 'Gak' signage   Signage system
Category: corporate architecture – communication media in architecture and public spaces

Manufacturer/Client
NAVER Corp.
South Korea

Design
NAVER Corp.
South Korea

NEC MultiSync EA294WMi   Monitor
Category: computer

Manufacturer/Client
NEC Display Solutions Ltd.
Japan

Design
NEC Display Solutions Ltd.
Japan

Nedap Smarttag Neck   Cow position and heat detection
Category: industry / skilled trades

Manufacturer/Client
Nedap Livestock Management
Netherlands

Design
WeLL Design
Netherlands

Manufacturer/Client **N**

Neo1 Smart pen
Category: computer

Manufacturer/Client
Neolab convergence
South Korea

Design
Neolab convergence
South Korea

MINI ME Coffee machine
Category: kitchen

Manufacturer/Client
Nescafé Dolce Gusto
Switzerland

Design
Multiple S. A. Global Design
Switzerland

new&able Website
Category: digital media – corporate websites

Manufacturer/Client
new&able
Germany

Design
hw.design GmbH
Germany

New Zealand Opera Brand
Category: print media – corporate design

Manufacturer/Client
New Zealand Opera
New Zealand

Design
Alt Group
New Zealand

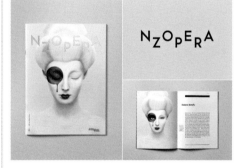

N'FERA SU1 Tire
Category: transportation design / special vehicles

Manufacturer/Client
NEXENTIRE Corporation
South Korea

Design
NEXENTIRE Corporation
South Korea

PLAY MUSEUM Communication architecture
Category: corporate architecture – communication media in architecture and public spaces

Manufacturer/Client
NHN Entertainment
South Korea

Design
NHN Entertainment
South Korea
NHN Entertainment Corp
South Korea
NAVER Corp.
South Korea

Manufacturer/Client **N**

Modern Life Makes Me Nervous — Book
Category: print media – publishing

Manufacturer/Client
Nico Bats, Fabian Bremer
Germany

Design
Nico Bats
Germany
Fabian Bremer
Germany

Nike National Team Kit — Football kit
Category: leisure / lifestyle

Manufacturer/Client
Nike, Inc.
United States of America

Design
Nike, Inc.
United States of America

Nike Studio Wrap — Studio workout shoe
Category: leisure / lifestyle

Manufacturer/Client
Nike, Inc.
United States of America

Design
Nike, Inc.
United States of America

GOLD

JURY STATEMENT

"What we like about this training shoe is that the design is reminiscent of the tradition and typology of a ballet shoe and thus communicates a kind of romantic feeling while being incredibly functional: the shoe provides great support although it still exposes parts of your foot. It also shows an intelligent use of the selected materials and works like a whole system where construction and material come together very sensitively."

Nike Flyknit Collective — A platform for innovators
Category: corporate architecture – installations in public spaces

Manufacturer/Client
Nike, Inc.
United States of America

Design
Nike, Inc.
United States of America

GOLD

JURY STATEMENT

"Nike has developed a platform for installations by creative innovators on which they can transmit the advantages of the Nike Flyknit material into physical structures and spaces within their own community. The means applied, the tools and also the size of the installations, which originated in London and Rio de Janeiro, among others, reflect the backgrounds of the respective artists. Cultural differences are perfectly staged."

Manufacturer/Client **N**

NIKE HIVIS 4SILO SHOWCASE Football shoes
Category: crossmedia – events

Manufacturer/Client
NIKE SPORTS KOREA CO., Ltd.
South Korea

Design
SILO Lab., South Korea
SID Design, South Korea
DOMY FACTORY, South Korea
NIKE SPORTS KOREA CO., Ltd.
South Korea

NIKON 1 AW1 Digital camera
Category: audio / video

Manufacturer/Client
Nikon Corporation
Japan

Design
Nikon Corporation
Japan

Pressed Chair Chair
Category: living room / bedroom

Manufacturer/Client
Nils Holger Moormann GmbH
Germany

Design
Harry Thaler
United Kingdom

COOLPIX A Digital camera
Category: audio / video

Manufacturer/Client
Nikon Corporation
Japan

Design
Nikon Corporation
Japan

AERO 21 / 26 Compact industrial vacuums
Category: industry / skilled trades

Manufacturer/Client
Nilfisk-Advance A / S
Denmark

Design
design-people
Denmark

iby6 Wheeled luggage
Category: leisure / lifestyle

Manufacturer/Client
Nilvia Lab
Italy

Design
Arch. Matteo Astolfi
Italy

Manufacturer/Client **N**

Slice Series LED armatures
Category: lighting

Manufacturer/Client
Ningbo Ledeshi Electric Equipment C
China

Design
npk design B. V.
Netherlands

Nissan Motor Show Motor show exhibition
Category: corporate architecture – installations in public spaces

Manufacturer/Client
Nissan Global
Japan

Design
George P Johnson
United States of America

No 2 Records Campaign
Category: print media – advertising media

Manufacturer/Client
No 2 Records
Germany

Design
Leo Burnett GmbH
Germany

itsumo kawaii Music promotion
Category: digital media – microsites

Manufacturer/Client
Nippon Columbia Co., Ltd.,
Japan

Design
Hakuhodo, Japan
METAPHOR Inc., Japan
PixelGrid Inc., Japan
ARMZ Inc., Japan

YONEJU Confiture Food packaging
Category: food

Manufacturer/Client
NISSHO JITSUGYO CO., Ltd.
Japan

Design
Toyo Seikan Group Holdings, Ltd.
Japan

Nokia Camera Mobile phone camera
Category: product interfaces

Manufacturer/Client
Nokia
Finland

Design
Nokia
Finland

Manufacturer/Client **N**

Nokia Lumia 1520 Smartphone
Category: telecommunications

Manufacturer/Client
Nokia
Finland

Design
Nokia
Finland

Nokia Lumia 520 Smartphone
Category: telecommunications

Manufacturer/Client
Nokia
China

Design
Nokia
China

Nokia Lumia 620 Smartphone
Category: telecommunications

Manufacturer/Client
Nokia
Finland

Design
Nokia
Finland

Nokia Lumia 625 Smartphone
Category: telecommunications

Manufacturer/Client
Nokia
China

Design
Nokia
China

Nokia Lumia 720 Smartphone
Category: telecommunications

Manufacturer/Client
Nokia
China

Design
Nokia
China

Nokia Lumia 2520 Tablet and power keyboard
Category: computer

Manufacturer/Client
Nokia
United States of America

Design
Nokia Design
United States of America

Manufacturer/Client **N**

Nokia CMD Portfolio Color and materials strategy
Category: material / textiles / wall+floor

Manufacturer/Client
Nokia
United Kingdom

Design
Nokia
United Kingdom

Nokia Lumia 1020 Mobile phone
Category: telecommunications

Manufacturer/Client
Nokia Corporation
Finland

Design
Nokia Corporation
Finland

Nokia Asha 501 Mobile phone
Category: telecommunications

Manufacturer/Client
Nokia Corporation
China

Design
Nokia Corporation
China

GOLD

JURY STATEMENT

"With the Asha 501, Nokia have done a fine job at producing a neat product that is well designed at a very reasonable price for the market. The colorful covers of the phone form a strong identity that resonates with its market. Furthermore, this product is executed very well."

Nokia New Asha Mobile phone
Category: telecommunications

Manufacturer/Client
Nokia Corporation
China

Design
Nokia Corporation
China

Ahoi Mechanical wrist watch
Category: leisure / lifestyle

Manufacturer/Client
NOMOS Glashütte / S. A.
Germany

Design
Berlinerblau GmbH
Germany

Manufacturer/Client N

concept plus Cookware
Category: household / tableware

Manufacturer/Client
Norbert Woll GmbH
Germany

Design
Friemel Design
Germany

CleanPot application App
Category: digital media – mobile applications

Manufacturer/Client
NOVAREVO
South Korea

Design
NOVAREVO
South Korea

Melicena Eiskrem Package design
Category: food

Manufacturer/Client
Novescor GmbH
Germany

Design
Leo Burnett GmbH
Germany

BS-3101 MI Base Station Base station
Category: telecommunications

Manufacturer/Client
NTT DOCOMO, Inc.
Japan

Design
Mitsubishi Electric Corporation
Japan

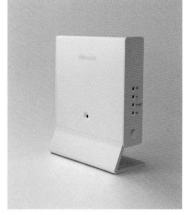

docomo Wireless Charger03 Wireless smartphone charger
Category: telecommunications

Manufacturer/Client
NTT DOCOMO, Inc.
Japan

Design
NTT DOCOMO, Inc.
Japan

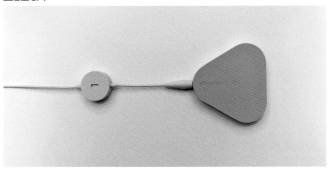

MITMACHEN.SPD Platform
Category: digital media – community / networking websites

Manufacturer/Client
NWMD GmbH
Germany

Design
3pc GmbH Neue Kommunikation
Germany

Manufacturer/Client **N**

Tex Glass® Textile glass
Category: material / textiles / wall+floor

Manufacturer/Client
Nya Nordiska Textiles GmbH
Germany

Design
Nya Nordiska Textiles GmbH
Germany

GOLD

JURY STATEMENT

"Tex Glass® is a decorative-laminated-glass-collection with an integrated fabric layer. It is an ideal space solution and an outstanding concept to give rooms an individual character. What convinced us with this design is that it presents entirely new possibilities in textile interior design and that the product may be applied in manifold ways."

Alicia Decoration fabric
Category: material / textiles / wall+floor

Manufacturer/Client
Nya Nordiska Textiles GmbH
Germany

Design
Nya Nordiska Textiles GmbH
Germany

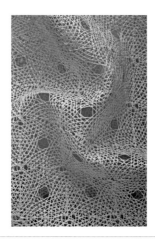

Quattro Multifunctional rail
Category: material / textiles / wall+floor

Manufacturer/Client
Nya Nordiska Textiles GmbH
Germany

Design
Nya Nordiska Textiles GmbH
Germany

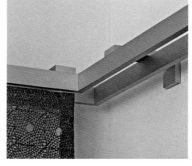

Viavai Decoration fabric
Category: material / textiles / wall+floor

Manufacturer/Client
Nya Nordiska Textiles GmbH
Germany

Design
Nya Nordiska Textiles GmbH
Germany

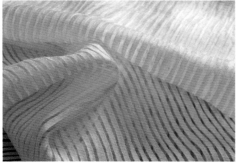

OAXIS Xrota N7 Audio speakers
Category: audio / video

Manufacturer/Client
OAXIS Holdings Pte. Ltd.
Singapore

Design
Gajah International Pte. Ltd.
Singapore

Manufacturer/Client **O**

Occhio Cases Book Project documentation
Category: print media – corporate design

Manufacturer/Client
Occhio GmbH
Germany

Design
Martin et Karczinski
Germany
Occhio GmbH
Germany

Occhio Preislisten Price lists
Category: print media – advertising media

Manufacturer/Client
Occhio GmbH
Germany

Design
Martin et Karczinski
Germany
Occhio GmbH
Germany

BLOB Mobile enclosures
Category: industry / skilled trades

Manufacturer/Client
Odenwälder Kunststoffwerke
Gehäusesysteme GmbH
Germany

Design
polyform Industrie Design
Germany

Ofa 3D Scan Non-contact measuring system
Category: medicine / health+care

Manufacturer/Client
Ofa Bamberg GmbH
Germany

Design
Andreas Ringelhan Industrialdesign
Germany

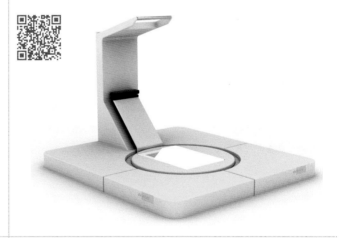

divo Drip stand
Category: medicine / health+care

Manufacturer/Client
Okamura Corporation
Japan

Design
Okamura Corporation
Japan
Medidea Corporation
Japan

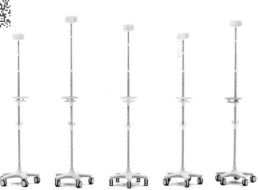

granite ware Natural humidifier
Category: research+development / professional concepts

Manufacturer/Client
Oki Seki Co., Ltd.
Japan

Design
IDL corporation
Japan

Manufacturer/Client **O**

GLANCE Luminaire family
Category: lighting

Manufacturer/Client
OLIGO Lichttechnik GmbH
Germany

Design
Ottenwälder und Ottenwälder
Germany

KITE Barber's chair
Category: public design

Manufacturer/Client
OLYMP GmbH & Co. KG
Germany

Design
code2design
Germany

Olympus OM-D E-M1 Mirror-less system camera
Category: audio / video

Manufacturer/Client
Olympus Corporation
Japan

Design
Olympus Corporation
Japan

Olympus PEN E-P5 Mirrorless system camera
Category: audio / video

Manufacturer/Client
Olympus Corporation
Japan

Design
Olympus Corporation
Japan

OLYMPUS OM-D Photography playground
Category: crossmedia – events

Manufacturer/Client
Olympus Deutschland GmbH
Germany

Design
vitamin e GmbH
Germany
eventlabs GmbH
Germany
flora&faunavisions GmbH
Germany

HT-B470 Electronic toothbrush
Category: bathroom / wellness

Manufacturer/Client
OMRON HEALTHCARE Co., Ltd.
Japan

Design
OMRON HEALTHCARE Co., Ltd.
Japan
C. Creative Inc.
Japan

Manufacturer/Client O

MC-520/521   Ear thermometer
Category: medicine / health+care

Manufacturer/Client
OMRON HEALTHCARE Co., Ltd.
Japan

Design
OMRON HEALTHCARE Co., Ltd.
Japan
YS design Inc.
Japan

MC-681   Digital thermometer
Category: medicine / health+care

Manufacturer/Client
OMRON HEALTHCARE Co., Ltd.
Japan

Design
OMRON HEALTHCARE Co., Ltd.
Japan
Design Studio S
Japan

OMRON Upper Arm series   Blood pressure monitor
Category: medicine / health+care

Manufacturer/Client
OMRON HEALTHCARE Co., Ltd.
Japan

Design
OMRON HEALTHCARE Co., Ltd.
Japan
Design Studio S
Japan

Nanchang Insun Cinema   Cinema
Category: corporate architecture – architecture / interior design

Manufacturer/Client
One Plus Partnership Ltd.
Hong Kong

Design
One Plus Partnership Ltd.
Hong Kong

Weldcap   Welding protection mask
Category: industry / skilled trades

Manufacturer/Client
Optrel AG
Switzerland

Design
Tribecraft AG
Switzerland

GOLD

JURY STATEMENT

"With this design of a welding-protection-mask, the LCD protection technique for the eyes was newly and innovatively designed. The lightness and flexibility in material ensures a completely new wearing comfort. Beyond the working safety, the focus also lies on the design quality. A functional, safe and visually appealing work tool with an entirely novel feeling of comfort has been designed – this is unprecedented."

Manufacturer/Client **O**

AYRTON On-ear headphones
Category: audio / video

Manufacturer/Client
Ora ïto Mobility
France

Design
Ora ïto Mobility
France

MICHA Travel charger
Category: computer

Manufacturer/Client
Ora ïto Mobility
France

Design
Ora ïto Mobility
France

ARKTIKA Pendant light
Category: lighting

Manufacturer/Client
Osram GmbH
Germany

Design
Osram GmbH
Germany

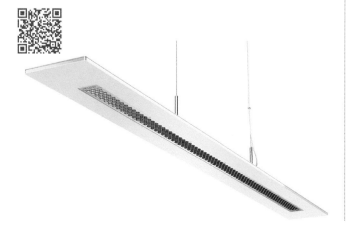

GIOTTO On-ear headphones
Category: audio / video

Manufacturer/Client
Ora ïto Mobility
France

Design
Ora ïto Mobility
France

HARU Brand identity design
Category: print media – corporate communication

Manufacturer/Client
ordinarypeople
South Korea

Design
ordinarypeople
South Korea

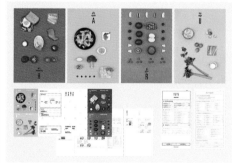

Structured Caos Shelter Bus shelter
Category: public design

Manufacturer/Client
Ótima
Brazil

Design
Indio da Costa AUDT
Brazil

Manufacturer/Client **O**

Voyager Evo Manual everyday wheelchair
Category: medicine / health+care

Manufacturer/Client
Otto Bock Mobility Solutions GmbH
Germany

Design
Otto Bock Mobility Solutions GmbH
Germany

GN 423 U-handles
Category: industry / skilled trades

Manufacturer/Client
Otto Ganter GmbH & Co. KG
Germany

Design
Otto Ganter GmbH & Co. KG
Germany

pinox Thermostat
Category: bathroom / wellness

Manufacturer/Client
Oventrop GmbH & Co. KG
Germany

Design
Oventrop GmbH & Co. KG
Germany
D&I
Germany

Personal Coach Sensor for tennis rackets
Category: leisure / lifestyle

Manufacturer/Client
Oxylane
France

Design
Oxylane
France

SWIP digital Watch
Category: leisure / lifestyle

Manufacturer/Client
Oxylane
France

Design
Oxylane
France

TR 800 Osmoz Adult tennis racket
Category: leisure / lifestyle

Manufacturer/Client
Oxylane
France

Design
Oxylane
France

Manufacturer/Client **O**

O!fun Miss Funtasy   Sex toys
Category: leisure / lifestyle

Manufacturer/Client
OZAKI INTERNATIONAL Co., Ltd.
Taiwan

Design
OZAKI INTERNATIONAL Co., Ltd.
Taiwan

Oz Estratégia+Design   Visual identity
Category: print media – corporate design

Manufacturer/Client
Oz Estratégia + Design
Brazil

Design
Oz Design Ltda.
Brazil

Aladdin Flip & Sip Mug   Beverage tumbler
Category: household / tableware

Manufacturer/Client
Pacific Market International
Netherlands

Design
Pacific Market International
Netherlands

O!arcade TAPiano   Game console
Category: computer

Manufacturer/Client
OZAKI INTERNATIONAL Co., Ltd.
Taiwan

Design
OZAKI INTERNATIONAL Co., Ltd.
Taiwan

Aladdin Crave Collection
Reusable lunch boxes and bottles
Category: household / tableware

Manufacturer/Client
Pacific Market International
Netherlands

Design
Pacific Market International
Netherlands

smart liner 240   Book-binding machine
Category: industry / skilled trades

Manufacturer/Client
Palamides GmbH
Germany

Design
defortec GmbH
Germany

Manufacturer/Client **P**

BQ-CC11 Rapid charger for batteries
Category: leisure / lifestyle

Manufacturer/Client
Panasonic Corporation
Japan

Design
Panasonic Corporation
Japan

DMC-GH3 Digital camera
Category: audio / video

Manufacturer/Client
Panasonic Corporation
Japan

Design
Panasonic Corporation
Japan

DMC-GM1 Digital camera
Category: audio / video

Manufacturer/Client
Panasonic Corporation
Japan

Design
Panasonic Corporation
Japan

DMC-GX7 Digital camera
Category: audio / video

Manufacturer/Client
Panasonic Corporation
Japan

Design
Panasonic Corporation
Japan

DMC-LF1 Digital camera
Category: audio / video

Manufacturer/Client
Panasonic Corporation
Japan

Design
Panasonic Corporation
Japan

SC-HTB170 Home cinema soundbar
Category: audio / video

Manufacturer/Client
Panasonic Corporation
Japan

Design
Panasonic Corporation
Japan

Manufacturer/Client **P**

SC-NA10   Wireless speaker
Category: audio / video

Manufacturer/Client
Panasonic Corporation
Japan

Design
Panasonic Corporation
Japan

TH-L55, L47WT60   HD liquid crystal television
Category: audio / video

Manufacturer/Client
Panasonic Corporation
Japan

Design
Panasonic Corporation
Japan

WV-SW598   Surveillance camera
Category: audio / video

Manufacturer/Client
Panasonic Corporation
Japan

Design
Panasonic Corporation
Japan

KX-PRS120   Cordless telephone
Category: telecommunications

Manufacturer/Client
Panasonic Corporation
Japan

Design
Panasonic Corporation
Japan

KX-PRW120   Cordless telephone
Category: telecommunications

Manufacturer/Client
Panasonic Corporation
Japan

Design
Panasonic Corporation
Japan

CF-AX2, CF-AX3   PC
Category: computer

Manufacturer/Client
Panasonic Corporation
Japan

Design
Panasonic Corporation
Japan

Manufacturer/Client **P**

BG-BL03   Solar lantern
Category: lighting

Manufacturer/Client
Panasonic Corporation
Japan

Design
Panasonic Corporation
Japan

FYY56020 series   LED office lighting
Category: lighting

Manufacturer/Client
Panasonic Corporation
Japan

Design
Panasonic Corporation
Japan

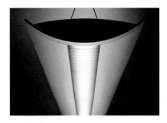

LightGuidePlate   LED lighting fixture
Category: lighting

Manufacturer/Client
Panasonic Corporation
Japan

Design
Panasonic Corporation
Japan

YYY16101 series   Outdoor lighting
Category: lighting

Manufacturer/Client
Panasonic Corporation
Japan

Design
Panasonic Corporation
Japan

NC-ZA1   Automatic espresso machine
Category: kitchen

Manufacturer/Client
Panasonic Corporation
Japan

Design
Panasonic Corporation
Japan

EW-DL82   Stain-off doltz
Category: bathroom / wellness

Manufacturer/Client
Panasonic Corporation
Japan

Design
Panasonic Corporation
Japan

Manufacturer/Client **P**

FV-47UD1 series Bathroom ventilator
Category: bathroom / wellness

Manufacturer/Client
Panasonic Corporation
Japan

Design
Panasonic Corporation
Japan

Panasonic Convention 2013 Event
Category: corporate architecture – exhibition / trade fair

Manufacturer/Client
Panasonic Deutschland
Germany

Design
BRAUNWAGNER GmbH
Germany
push design
Germany

VEGA Secret Note Smartphone packaging
Category: consumer electronics

Manufacturer/Client
PANTECH CO., Ltd.
South Korea

Design
PANTECH CO., Ltd.
South Korea

Paper Pleasure Corporate Design
Category: print media – corporate design

Manufacturer/Client
Paper Pleasure
Germany

Design
Clormann Design GmbH
Germany

Parkett Design Edition Engineered wood flooring
Category: material / textiles / wall+floor

Manufacturer/Client
Parador GmbH & Co. KG
Germany

Design
Hadi Teherani AG
Germany

CyFlow® Cube 6 Cell analysis system
Category: medicine / health+care

Manufacturer/Client
Partec GmbH
Germany

Design
formfreun.de
gestaltungsgesellschaft
Germany

Manufacturer/Client **P**

CyFox® Medical laboratory
Category: medicine / health+care

Manufacturer/Client
Partec GmbH
Germany

Design
formfreun.de
gestaltungsgesellschaft
Germany

D800 Self-service point-of-sale terminal
Category: public design

Manufacturer/Client
PAX Computer Technology Co., Ltd.
China

Design
TGS Design Consultancy
China

Thor Wireless Touch Display
Category: computer

Manufacturer/Client
Pegatron Corporation
Taiwan

Design
PEGA D&E
Taiwan

Flux Intelligence Concept video
Category: research+development /professional concepts

Manufacturer/Client
Pegatron Corporation
Taiwan

Design
PEGA D&E
Taiwan

Inshell Door lever and door stopper
Category: buildings

Manufacturer/Client
Pegatron Corporation
Taiwan

Design
PEGACASA Design Team
Taiwan

pester pac automation Deep moulded papertray
Category: research+development / professional concepts

Manufacturer/Client
pester pac automation GmbH
Germany

Design
designship
Germany
TU Dresden
Germany

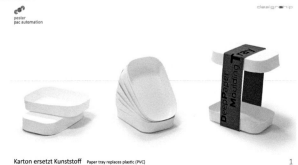

Karton ersetzt Kunststoff Paper tray replaces plastic (PVC)

Manufacturer/Client **P**

Business of a Lifetime Business report box
Category: print media – advertising media

Manufacturer/Client
Peyer Graphic GmbH, Germany
Ernst A. Geese GmbH, Germany
Kösel GmbH & Co. KG, Germany

Design
JUNO
Germany

Elevate Caledon jacket Jacket
Category: leisure / lifestyle

Manufacturer/Client
PF Concept International B. V.
Netherlands

Design
PF Concept International B. V.
Netherlands

Marksman Carve Ballpen
Category: office / business

Manufacturer/Client
PF Concept International B. V.
Netherlands

Design
PF Concept International B. V.
Netherlands

Avenue Flow bottle Drinking bottle
Category: leisure / lifestyle

Manufacturer/Client
PF Concept International B. V.
Netherlands

Design
PF Concept International B. V.
Netherlands

Elevate Mani Power Fleece Fleece jacket
Category: leisure / lifestyle

Manufacturer/Client
PF Concept International B. V.
Netherlands

Design
PF Concept International B. V.
Netherlands

Marksman Explorer Ballpen
Category: office / business

Manufacturer/Client
PF Concept International B. V.
Netherlands

Design
PF Concept International B. V.
Netherlands

Manufacturer/Client **P**

Avenue Flow tumbler   Isolating tumbler
Category: household / tableware

Manufacturer/Client
PF Concept International B. V.
Netherlands

Design
PF Concept International B. V.
Netherlands

Pfeffersack & Soehne   Vanilla packaging
Category: food

Manufacturer/Client
Pfeffersack & Soehne
Germany

Design
Pfeffersack & Soehne
Germany

blueglobe CLEAN® Plus   Cable gland
Category: industry / skilled trades

Manufacturer/Client
PFLITSCH GmbH & Co. KG
Germany

Design
PFLITSCH GmbH & Co. KG
Germany

Phiaton Bridge   Headphones
Category: audio / video

Manufacturer/Client
Phiaton
United States of America

Design
TEAGUE
United States of America

CASCARA   Nut bowl with depot
Category: household / tableware

Manufacturer/Client
PHILIPPI GmbH
Germany

Design
Murken Hansen Product Design
Germany

Gmate™ SMART   Blood glucose monitoring system
Category: medicine / health+care

Manufacturer/Client
philosys Co., Ltd.
South Korea

Design
philosys Co., Ltd.
South Korea

233

Manufacturer/Client **P**

Axon Tapware
Category: bathroom / wellness

Manufacturer/Client
Phoenix Tapware
Australia

Design
Phoenix Tapware
Australia

Ortho Faucet
Category: bathroom / wellness

Manufacturer/Client
Phoenix Tapware
Australia

Design
Phoenix Tapware
Australia

VISOR Marker
Category: office / business

Manufacturer/Client
Pica-Marker
Germany

Design
winkelbauer-design
Germany

Kawa Tapware
Category: bathroom / wellness

Manufacturer/Client
Phoenix Tapware
Australia

Design
Phoenix Tapware
Australia

PVP Communication campaign
Category: crossmedia – advertising / campaigns

Manufacturer/Client
Piano Viabilità Polo Luganese
Switzerland

Design
CCRZ sa
Switzerland
comunicAzione
Switzerland
Agenzia di Comunicazione
Switzerland

RIG Gaming headset
Category: audio / video

Manufacturer/Client
Plantronics
United States of America

Design
Plantronics Design
United States of America

Manufacturer/Client **P**

Voyager Legend CS Bluetooth headset
Category: telecommunications

Manufacturer/Client
Plantronics
United States of America

Design
Plantronics Design
United States of America

P+US ZERO 2 Eyewear
Category: leisure / lifestyle

Manufacturer/Client
Plus Eyewear Limited
Hong Kong

Design
Plus Eyewear Limited
Hong Kong

Hygge 3012 Watch series
Category: leisure / lifestyle

Manufacturer/Client
P.O.S. Co., Ltd.
Japan

Design
Pentagon Design Ltd.
Finland

WORIGAWI Wine packaging
Category: packaging formdesign

Manufacturer/Client
Plateau Wine Trading
South Korea

Design
Manifesto Architecture P. C.
United States of America

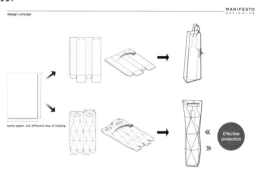

POLA B.A MEN Cosmetics
Category: beauty / health / household

Manufacturer/Client
POLA Inc.
Japan

Design
POLA Inc.
Japan

PROBAT shop roaster Product brochure
Category: print media – product communication

Manufacturer/Client
PROBAT-Werke von Gimborn
Maschinenfabrik GmbH
Germany

Design
Agentur romen
Germany

Manufacturer/Client **P**

Promate bluTrend Bluetooth headphone
Category: audio / video

Manufacturer/Client
Promate Technologies
China

Design
Promate Technologies
China

Promate Ovally Lightning USB powerbank
Category: telecommunications

Manufacturer/Client
Promate Technologies
China

Design
Promate Technologies
China

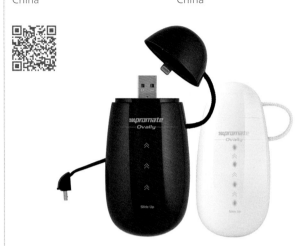

P & T – Paper & Tea Tea packaging
Category: beverages

Manufacturer/Client
P & T – Paper & Tea
Germany

Design
Sonnenstaub –
Büro für Gestaltung und Illustration
Germany

AMIO Food packaging
Category: food

Manufacturer/Client
Pulmuone Health & Living Co., Ltd.
South Korea

Design
cd's associates
South Korea

Qin Sheng Breeze Tea gift set
Category: beverages

Manufacturer/Client
Qinhetang International Co., Ltd.
Taiwan

Design
Existence Design Co., Ltd.
Taiwan

Qisda QCM-501 Smartphone
Category: telecommunications

Manufacturer/Client
Qisda Corporation
Taiwan

Design
Qisda Creative Design Center
Taiwan

Manufacturer/Client Q

Qisda QTD-504 Monitor
Category: computer

Manufacturer/Client
Qisda Corporation
Taiwan

Design
Qisda Creative Design Center
Taiwan

QNAP Silent NAS Series Network attached storage
Category: computer

Manufacturer/Client
QNAP Systems, Inc.
Taiwan

Design
QNAP Systems, Inc.
Taiwan

QDH 4G LTE customer premise equipment
Category: telecommunications

Manufacturer/Client
Quanta Computer Inc.
Taiwan

Design
Quanta Computer Inc.
Taiwan

QUECHUA Softshell Spread Man Softshell jacket
Category: leisure / lifestyle

Manufacturer/Client
QUECHUA
France

Design
QUECHUA
France

Among the Mountains Sofa
Category: living room / bedroom

Manufacturer/Client
QUMEI FURNITURE GROUP Co., Ltd.
China

Design
YUAN YUAN STUDIO
China

Pea Princess Lounge chair
Category: living room / bedroom

Manufacturer/Client
QUMEI FURNITURE GROUP Co., Ltd.
China

Design
YUAN YUAN STUDIO
China

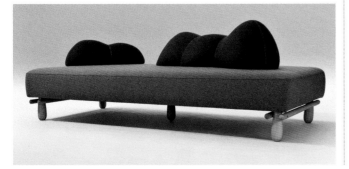

Manufacturer/Client **R**

Stadtlounge St. Gallen Public living room
Category: corporate architecture – installations in public spaces

Manufacturer/Client
Raiffeisenbank
Switzerland

Design
Carlos Martinez Architekten AG
Switzerland
Pipilotti Rist
Switzerland

Sophie Sofa
Category: living room / bedroom

Manufacturer/Client
Raum B Architektur
Switzerland

Design
Raum B Architektur
Switzerland

Vanish Gel Bottle range
Category: beauty / health / household

Manufacturer/Client
Reckitt Benckiser
United Kingdom

Design
VanBerlo B. V.
Netherlands
Lothar Böhm Associates
Germany

Engine Room Cookbook Book
Category: print media – publishing

Manufacturer/Client
Random House
New Zealand

Design
Alt Group
New Zealand

R5 3D light field camera
Category: industry / skilled trades

Manufacturer/Client
Raytrix GmbH
Germany

Design
Designfactor
Germany

rampreport – Design Magazine
Category: print media – publishing

Manufacturer/Client
Red Indians Publishing
GmbH & Co. KG
Germany

Design
Seidldesign
Germany

Manufacturer/Client **R**

rampstyle No.5 – Bad Boys   Magazine
Category: print media – publishing

Manufacturer/Client
Red Indians Publishing
GmbH & Co. KG
Germany

Design
Seidldesign
Germany

Blossom cutlery   Plastic cutlery
Category: household / tableware

Manufacturer/Client
Reflect Inc.
South Korea

Design
Reflect Inc.
South Korea

Carry lumiinary   Humidifier
Category: research+development / professional concepts

Manufacturer/Client
Reiko Kitora
Japan

Design
Atsuhito Kitora
Japan

Cradle of the forest   Wires and eco-hooks
Category: research+development / professional concepts

Manufacturer/Client
Reiko Kitora
Japan

Design
Atsuhito Kitora
Japan

COGOO Terminal   Bicycle rental terminal
Category: public design

Manufacturer/Client
Relations Inc.
Japan

Design
Relations Inc., Japan
TBWAHAKUHODO
HAKUHODO Inc., Japan
ziba tokyo Co., Ltd., Japan
KENJIRO NII AND DESIGN, Japan

COBOGÓ ATOLL   Wall tile
Category: material / textiles / wall+floor

Manufacturer/Client
Renata Rubim
Brazil

Design
Renata Rubim
Brazil

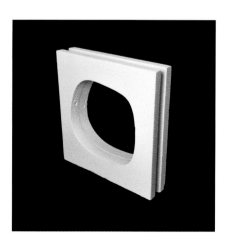

Manufacturer/Client **R**

Cov-Box Desktop charge station
Category: telecommunications

Manufacturer/Client
RenQing Technology Co., Ltd.
China

Design
RenQing Technology Co., Ltd.
China

Delta Mobile power bank
Category: telecommunications

Manufacturer/Client
RenQing Technology Co., Ltd.
China

Design
RenQing Technology Co., Ltd.
China

Voi Bluetooth headset and power bank
Category: telecommunications

Manufacturer/Client
RenQing Technology Co., Ltd.
China

Design
RenQing Technology Co., Ltd.
China

Atlanta Steam bath
Category: bathroom / wellness

Manufacturer/Client
repaBAD GmbH
Germany

Design
repaBAD GmbH
Germany

REWE – Besser leben iOs / Android App
Category: digital media – mobile applications

Manufacturer/Client
REWE Markt GmbH
Germany

Design
SapientNitro
Germany

RICOH THETA Spherical camera
Category: product interfaces

Manufacturer/Client
RICOH
Japan

Design
RICOH, Japan
TRIAND Inc., Japan

Manufacturer/Client **R**

GR Digital camera
Category: audio / video

Manufacturer/Client
RICOH
Japan

Design
RICOH
Japan

RICOH Pro 8100 Series Production printers
Category: computer

Manufacturer/Client
RICOH
Japan

Design
RICOH
Japan

BOSCH KMF40S20TI Multidoor refrigerator
Category: kitchen

Manufacturer/Client
Robert Bosch Hausgeräte GmbH
Germany

Design
Robert Bosch Hausgeräte GmbH
Germany

GOLD

JURY STATEMENT

"This fridge from Bosch is beautifully proportioned with good size drawers. The choice of materials is excellent and the engineering quality is fantastic. It has a nice and easy-to-use system inside of the fridge. This is simply an outstanding product."

BOSCH DWK09G660/.-620 Chimney hood
Category: kitchen

Manufacturer/Client
Robert Bosch Hausgeräte GmbH
Germany

Design
Robert Bosch Hausgeräte GmbH
Germany

BOSCH HBG78B960 A-30% Built-in oven, vulcano black
Category: kitchen

Manufacturer/Client
Robert Bosch Hausgeräte GmbH
Germany

Design
Robert Bosch Hausgeräte GmbH
Germany

Manufacturer/Client **R**

BOSCH KGD36VI30   Fridge-freezer-combination
Category: kitchen

Manufacturer/Client
Robert Bosch Hausgeräte GmbH
Germany

Design
Robert Bosch Hausgeräte GmbH
Germany

BOSCH KGV36VH30S   Colored fridge-freezer
Category: kitchen

Manufacturer/Client
Robert Bosch Hausgeräte GmbH
Germany

Design
Robert Bosch Hausgeräte GmbH
Germany

BOSCH PIU875K17E   Induction hob
Category: kitchen

Manufacturer/Client
Robert Bosch Hausgeräte GmbH
Germany

Design
Robert Bosch Hausgeräte GmbH
Germany

BOSCH PPS916B91E   Tempered glass gas cooktop
Category: kitchen

Manufacturer/Client
Robert Bosch Hausgeräte GmbH
Germany

Design
Robert Bosch Hausgeräte GmbH
Germany

BOSCH SBE 69U11EU   Fully integrated dishwasher
Category: kitchen

Manufacturer/Client
Robert Bosch Hausgeräte GmbH
Germany

Design
Robert Bosch Hausgeräte GmbH
Germany

BOSCH SMZ 5300   Glass secure tray
Category: kitchen

Manufacturer/Client
Robert Bosch Hausgeräte GmbH
Germany

Design
Robert Bosch Hausgeräte GmbH
Germany

Manufacturer/Client **R**

GSD36PI20 / KSW36PI30   Twincenter
Category: kitchen

Manufacturer/Client
Robert Bosch Hausgeräte GmbH
Germany

Design
Robert Bosch Hausgeräte GmbH
Germany

MFQ36 Serie   Hand mixer
Category: kitchen

Manufacturer/Client
Robert Bosch Hausgeräte GmbH
Germany

Design
Robert Bosch Hausgeräte GmbH
Germany

Tassimo VIVY   Multi beverage maker
Category: kitchen

Manufacturer/Client
Robert Bosch Hausgeräte GmbH
Germany

Design
Robert Bosch Hausgeräte GmbH
Germany
pearl creative
Germany

BCH6 Serie   Bagless vacuum cleaner
Category: household / tableware

Manufacturer/Client
Robert Bosch Hausgeräte GmbH
Germany

Design
Robert Bosch Hausgeräte GmbH
Germany
BRANDIS Industrial Design
Germany

BOSCH WAY287W3   Washing machine
Category: household / tableware

Manufacturer/Client
Robert Bosch Hausgeräte GmbH
Germany

Design
Robert Bosch Hausgeräte GmbH
Germany

TDA70 Serie   Steam iron
Category: household / tableware

Manufacturer/Client
Robert Bosch Hausgeräte GmbH
Germany

Design
Robert Bosch Hausgeräte GmbH
Germany

Manufacturer/Client **R**

Drift jugs range Jugs
Category: household / tableware

Manufacturer/Client
Robert Welch Designs Ltd.
United Kingdom

Design
Robert Welch Designs Ltd.
United Kingdom

DIE SCHOENHEIT Magalogue: magazine catalogue
Category: print media – product communication

Manufacturer/Client
rosconi GmbH
Germany

Design
atelier schneeweiss GmbH
Germany

TAKUSHI Table
Category: living room / bedroom

Manufacturer/Client
Röthlisberger Kollektion
Switzerland

Design
Gavin Harris
Australia

AVENT Comfort Electrical breast pumps
Category: leisure / lifestyle

Manufacturer/Client
Royal Philips Electronics
Netherlands

Design
Philips Design
Netherlands

DesignLine PDL8908S/12 LED TV
Category: audio / video

Manufacturer/Client
Royal Philips Electronics
Netherlands

Design
Philips Design
Netherlands

GOLD

JURY STATEMENT

"This television represents an extraordinary product and design achievement. It impresses through its casual gesture of simply being leant against the wall. No eye-catching detail stands out in any way. In principle, it is simply a plain pane of glass that only comes to life when it is being switched on."

Manufacturer/Client **R**

Metronomis LED range LED outdoor lighting
Category: lighting

Manufacturer/Client
Royal Philips Electronics
Netherlands

Design
Philips Design
Netherlands

GOLD

JURY STATEMENT

"The sensation of the materials in this LED outdoor lighting is fantastic. The customizable options that can be used to create different patterns really opens the door to countless creative possibilities, which is especially exciting for designers and architects."

Sonicare DiamondClean Power toothbrush
Category: bathroom / wellness

Manufacturer/Client
Royal Philips Electronics
Netherlands

Design
Philips Design
Netherlands

GOLD

JURY STATEMENT

"The electric toothbrush Sonicare DiamondClean Black innovatively combines the most modern technology with an intuitive user interface and the utmost in hygiene: the user chooses the desired type of cleaning with just one finger, the device is pleasant to hold and the excellently manufactured casing is easily cleaned. The timeless and elegant design also impressed us."

Lifeline GoSafe, HomeSafe Emergency help communicators
Category: medicine / health+care

Manufacturer/Client
Royal Philips Electronics
Netherlands

Design
Philips Design
Netherlands

GOLD

JURY STATEMENT

"We think this immediate-help-communicator is extremely well made as it combines automatic fall recognition with emergency call technology and location transmission. The transmitter features an adaptable design and is visually appealing. It is an innovative design which contributes toward optimal support and care within society."

Manufacturer/Client **R**

Wake-up Light HF Natural alarm clock
Category: leisure / lifestyle

Manufacturer/Client
Royal Philips Electronics
Netherlands

Design
Philips Design
Netherlands

7000 Range 47PFL7108S/12 LED TV
Category: audio / video

Manufacturer/Client
Royal Philips Electronics
Netherlands

Design
Philips Design
Netherlands

BR-1X SB5100/SB5200 Portable Bluetooth speaker
Category: audio / video

Manufacturer/Client
Royal Philips Electronics
Netherlands

Design
Philips Design
Netherlands

Fidelio Home Cinema CSS7235Y (E5)
Home cinema sound system
Category: audio / video

Manufacturer/Client
Royal Philips Electronics
Netherlands

Design
Philips Design
Netherlands

Fidelio L2 Headphone
Category: audio / video

Manufacturer/Client
Royal Philips Electronics
Netherlands

Design
Philips Design
Netherlands

Fidelio P9x + P8 Wireless portable speakers
Category: audio / video

Manufacturer/Client
Royal Philips Electronics
Netherlands

Design
Philips Design
Netherlands

Manufacturer/Client **R**

Fidelio S2 (S2BK/S2WT)   Headphone
Category: audio / video

Manufacturer/Client
Royal Philips Electronics
Netherlands

Design
Philips Design
Netherlands

InSight M120 home monitor   Home monitor
Category: audio / video

Manufacturer/Client
Royal Philips Electronics
Netherlands

Design
Philips Design
Netherlands

M1X-DJ – DJ Sound System DS8900   Sound system
Category: audio / video

Manufacturer/Client
Royal Philips Electronics
Netherlands

Design
Philips Design
Netherlands

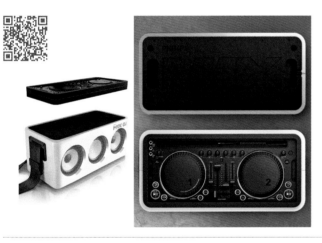

Screeneo   Smart LED Projector
Category: audio / video

Manufacturer/Client
Royal Philips Electronics
Netherlands

Design
Philips Design
Netherlands

Splash proof SB2000 DOT   Wireless portable speaker
Category: audio / video

Manufacturer/Client
Royal Philips Electronics
Netherlands

Design
Philips Design
Netherlands

GentleSpace Gen-2   Luminiare
Category: lighting

Manufacturer/Client
Royal Philips Electronics
Netherlands

Design
Philips Design
Netherlands

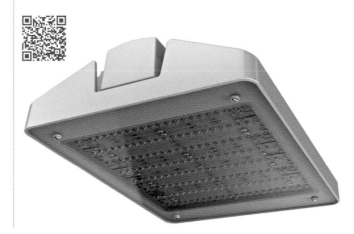

Manufacturer/Client **R**

hue bulb LED bulb
Category: lighting

Manufacturer/Client
Royal Philips Electronics
Netherlands

Design
Philips Design
Netherlands

hue personal wireless Lighting system
Category: lighting

Manufacturer/Client
Royal Philips Electronics
Netherlands

Design
Philips Design
Netherlands

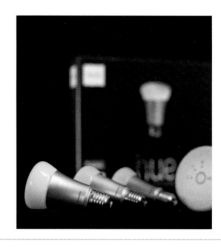

Lirio by Philips-La Lente LED table lamps
Category: lighting

Manufacturer/Client
Royal Philips Electronics
Netherlands

Design
Philips Design
Netherlands

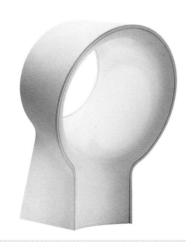

Lirio by Philips-Piculet Table and reading lamp
Category: lighting

Manufacturer/Client
Royal Philips Electronics
Netherlands

Design
Philips Design
Netherlands

Lirio Philips-Piega Luce Pendants and wall lamps
Category: lighting

Manufacturer/Client
Royal Philips Electronics
Netherlands

Design
Philips Design
Netherlands

MyGarden Solar-Dusk family Outdoor light fixtures
Category: lighting

Manufacturer/Client
Royal Philips Electronics
Netherlands

Design
Philips Design
Netherlands

Manufacturer/Client **R**

SmartBalance suspended   Suspended LED luminaire
Category: lighting

Manufacturer/Client
Royal Philips Electronics
Netherlands

Design
Philips Design
Netherlands

Avance Collection Airfryer XL   Frying with air
Category: kitchen

Manufacturer/Client
Royal Philips Electronics
Netherlands

Design
Philips Design
Netherlands

Café Gourmet HD5412   Coffee maker
Category: kitchen

Manufacturer/Client
Royal Philips Electronics
Netherlands

Design
Philips Design
Netherlands

SENSEO® Milk Twister   Milk twister
Category: kitchen

Manufacturer/Client
Royal Philips Electronics
Netherlands

Design
Philips Design
Netherlands

AC4081 / AC4080   Air purifier and humidifier
Category: household / tableware

Manufacturer/Client
Royal Philips Electronics
Netherlands

Design
Philips Design
Netherlands

Desktop Humidifier HU4706   Air humidifier
Category: household / tableware

Manufacturer/Client
Royal Philips Electronics
Netherlands

Design
Philips Design
Netherlands

Manufacturer/Client **R**

EasyPro FC-Serie   Bagless vacuum cleaner
Category: household / tableware

Manufacturer/Client   **Design**
Royal Philips Electronics   Philips Design
Netherlands   Netherlands

PerfectCare Pure GC7635   Pressurized steam generator
Category: household / tableware

Manufacturer/Client   **Design**
Royal Philips Electronics   Philips Design
Netherlands   Netherlands

Beard Trimmer 9000   Beard trimmer
Category: bathroom / wellness

Manufacturer/Client   **Design**
Royal Philips Electronics   Philips Design
Netherlands   Netherlands

Electric Shaver for China   Shaver
Category: bathroom / wellness

Manufacturer/Client   **Design**
Royal Philips Electronics   Philips Design
Netherlands   Netherlands

Sonicare FlexCare Platinum   Power toothbrush
Category: bathroom / wellness

Manufacturer/Client   **Design**
Royal Philips Electronics   Philips Design
Netherlands   Netherlands

Sonicare PowerUp battery   Battery toothbrush
Category: bathroom / wellness

Manufacturer/Client   **Design**
Royal Philips Electronics   Philips Design
Netherlands   Netherlands

Manufacturer/Client **R**

Healthy Heroes   Educational interactive game
Category: medicine / health+care

Manufacturer/Client
Royal Philips Electronics
Netherlands

Design
Philips Design
Netherlands

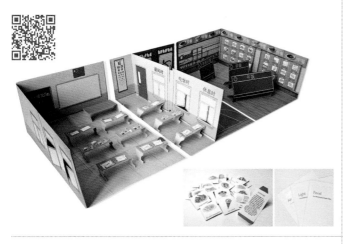

IBA Proton Therapy Suite   IBA proton therapy room
Category: medicine / health+care

Manufacturer/Client
Royal Philips Electronics
Netherlands

Design
Philips Design
Netherlands

IQon Spectral   CT scanner
Category: medicine / health+care

Manufacturer/Client
Royal Philips Electronics
Netherlands

Design
Philips Design
Netherlands

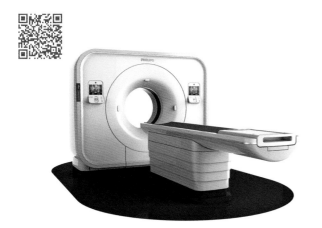

Multiva   MR system
Category: medicine / health+care

Manufacturer/Client
Royal Philips Electronics
Netherlands

Design
Philips Design
Netherlands

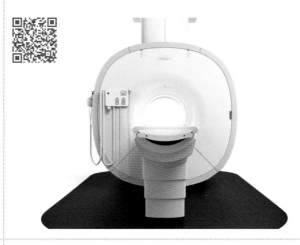

Nuance & Nuance Pro   Mask system
Category: medicine / health+care

Manufacturer/Client
Royal Philips Electronics
Netherlands

Design
Philips Design
Netherlands

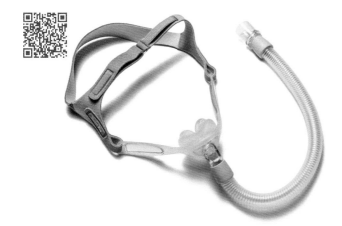

Vereos   Digital PET CT scanner
Category: medicine / health+care

Manufacturer/Client
Royal Philips Electronics
Netherlands

Design
Philips Design
Netherlands

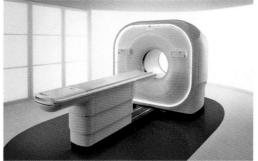

Manufacturer/Client **R**

VISIQ Portable ultrasound system
Category: medicine / health+care

Manufacturer/Client
Royal Philips Electronics
Netherlands

Design
Philips Design
Netherlands

Delivery experience concept Delivery experience concept
Category: research+development / professional concepts

Manufacturer/Client
Royal Philips Electronics
Netherlands

Design
Philips Design
Netherlands

Ergon Branddesign
Category: print media – corporate design

Manufacturer/Client
RTI Sports GmbH
Germany

Design
KW43 BRANDDESIGN
Germany

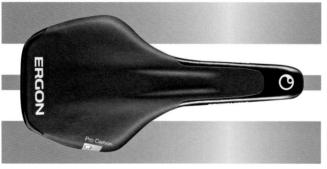

mono architecture Remote control
Category: audio / video

Manufacturer/Client
ruwido austria GmbH
Austria

Design
Zeug Design GmbH
Austria

Wall Mounted speaker Bluetooth speaker
Category: audio / video

Manufacturer/Client
Ryohin Keikaku Co., Ltd.
Japan

Design
NAOTO FUKASAWA DESIGN
Japan

USB Swing Desk fan Desk fan
Category: office / business

Manufacturer/Client
Ryohin Keikaku Co., Ltd.
Japan

Design
James Irvine S. R. L.
Italy

Manufacturer/Client **R**

Reclining sofa Reclining sofa
Category: living room / bedroom

Manufacturer/Client
Ryohin Keikaku Co., Ltd.
Japan

Design
Ryohin Keikaku Co., Ltd., Japan
MIYAKE Design Office, Japan

MUJI Air Circulator Air circulator
Category: household / tableware

Manufacturer/Client
Ryohin Keikaku Co., Ltd.
Japan

Design
MIYAKE Design Office
Japan

weber.therm style Glas ETICS with design glass surface
Category: buildings

Manufacturer/Client
Saint-Gobain Weber GmbH
Germany

Design
Saint-Gobain Weber GmbH
Germany

223 plus Customer magazine
Category: print media – corporate communication

Manufacturer/Client
Sal. Oppenheim jr. & Cie. KGaA
Germany

Design
Simon & Goetz Design
GmbH & Co. KG
Germany

Agreste Rug
Category: material / textiles / wall+floor

Manufacturer/Client
Salvatore Tiles & Tapetes
Brazil

Design
Salvatore Tiles & Tapetes
Brazil

ela Decorative hinge cover
Category: living room / bedroom

Manufacturer/Client
Samet
Turkey

Design
KozSusani Design
United States of America

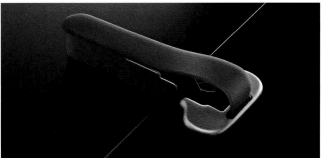

Manufacturer/Client **S**

Glasslock-Pure Glass container packaging
Category: beauty / health / household

Manufacturer/Client
Samkwang Glass Co., Ltd.
South Korea

Design
Samkwang Glass Co., Ltd.
South Korea

Block Canister Canister
Category: household / tableware

Manufacturer/Client
Samkwang Glass Co., Ltd.
South Korea

Design
Samkwang Glass Co., Ltd.
South Korea

Glasslock-smart Glass container
Category: household / tableware

Manufacturer/Client
Samkwang Glass Co., Ltd.
South Korea

Design
Samkwang Glass Co., Ltd.
South Korea

In Full Swing Corporate Report 2012
Category: print media – corporate communication

Manufacturer/Client
Samsung Cheil Industries
South Korea

Design
Talantone Creative Group
South Korea

BUMBLEBEE Smartphone accessory
Category: research+development / professional concepts

Manufacturer/Client
Samsung Electronics
South Korea

Design
Samsung Electronics
South Korea

Contents-oriented Control Remote controller concept
Category: research+development / professional concepts

Manufacturer/Client
Samsung Electronics
South Korea

Design
Samsung Electronics
South Korea

Manufacturer/Client **S**

2013 SMART TV UX   TV
Category: product interfaces

Manufacturer/Client
Samsung Electronics Co., Ltd.
South Korea

Design
Samsung Electronics Co., Ltd.
South Korea

Samsung WatchON   Tablet App
Category: digital media – mobile applications

Manufacturer/Client
Samsung Electronics Co., Ltd.
South Korea

Design
Samsung Electronics Co., Ltd.
South Korea

Explore 3D   3D television
Category: product interfaces

Manufacturer/Client
Samsung Electronics Co., Ltd.
South Korea

Design
Samsung Electronics Co., Ltd.
South Korea

F9000   UHD TV
Category: audio / video

Manufacturer/Client
Samsung Electronics Co., Ltd.
South Korea

Design
Samsung Electronics Co., Ltd.
South Korea

GALAXY NX   Digital camera
Category: audio / video

Manufacturer/Client
Samsung Electronics Co., Ltd.
South Korea

Design
Samsung Electronics Co., Ltd.
South Korea

HW-F750 / 550 Series   Soundbar
Category: audio / video

Manufacturer/Client
Samsung Electronics Co., Ltd.
South Korea

Design
Samsung Electronics Co., Ltd.
South Korea

Manufacturer/Client **S**

KN55S9C OLED TV
Category: audio / video

Manufacturer/Client
Samsung Electronics Co., Ltd.
South Korea

Design
Samsung Electronics Co., Ltd.
South Korea

NX300 Digital camera
Category: audio / video

Manufacturer/Client
Samsung Electronics Co., Ltd.
South Korea

Design
Samsung Electronics Co., Ltd.
South Korea

One Connect TV external jack pack
Category: audio / video

Manufacturer/Client
Samsung Electronics Co., Ltd.
South Korea

Design
Samsung Electronics Co., Ltd.
South Korea

Smart Input Device Remote control and keyboard
Category: audio / video

Manufacturer/Client
Samsung Electronics Co., Ltd.
South Korea

Design
Samsung Electronics Co., Ltd.
South Korea

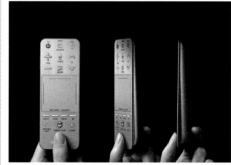

UN85S9 UHD TV
Category: audio / video

Manufacturer/Client
Samsung Electronics Co., Ltd.
South Korea

Design
Samsung Electronics Co., Ltd.
South Korea

Galaxy Gear Watch
Category: telecommunications

Manufacturer/Client
Samsung Electronics Co., Ltd.
South Korea

Design
Samsung Electronics Co., Ltd.
South Korea

Manufacturer/Client **S**

Galaxy S4   Smartphone
Category: telecommunications

Manufacturer/Client
Samsung Electronics Co., Ltd.
South Korea

Design
Samsung Electronics Co., Ltd.
South Korea

Note3 S View Cover   Safety case
Category: telecommunications

Manufacturer/Client
Samsung Electronics Co., Ltd.
South Korea

Design
Samsung Electronics Co., Ltd.
South Korea

Ativ Book9 Plus   Notebook
Category: computer

Manufacturer/Client
Samsung Electronics Co., Ltd.
South Korea

Design
Samsung Electronics Co., Ltd.
South Korea

M2020 Series   Compact printer series
Category: computer

Manufacturer/Client
Samsung Electronics Co., Ltd.
South Korea

Design
Samsung Electronics Co., Ltd.
South Korea

M2825 / M2875 Series   Efficient mono printer
Category: computer

Manufacturer/Client
Samsung Electronics Co., Ltd.
South Korea

Design
Samsung Electronics Co., Ltd.
South Korea

M4020 / M4070 Series   Performance mono printer
Category: computer

Manufacturer/Client
Samsung Electronics Co., Ltd.
South Korea

Design
Samsung Electronics Co., Ltd.
South Korea

Manufacturer/Client **S**

Mate Mono laser printer
Category: computer

Manufacturer/Client
Samsung Electronics Co., Ltd.
South Korea

Design
Samsung Electronics Co., Ltd.
South Korea

NL22B Showcase
Category: computer

Manufacturer/Client
Samsung Electronics Co., Ltd.
South Korea

Design
Samsung Electronics Co., Ltd.
South Korea

One and One Dual color laser printer
Category: computer

Manufacturer/Client
Samsung Electronics Co., Ltd.
South Korea

Design
Samsung Electronics Co., Ltd.
South Korea

SB970 PLS Monitor Monitor
Category: computer

Manufacturer/Client
Samsung Electronics Co., Ltd.
South Korea

Design
Samsung Electronics Co., Ltd.
South Korea

SC750 Monitor
Category: computer

Manufacturer/Client
Samsung Electronics Co., Ltd.
South Korea

Design
Samsung Electronics Co., Ltd.
South Korea

SC770 Touch monitor Monitor
Category: computer

Manufacturer/Client
Samsung Electronics Co., Ltd.
South Korea

Design
Samsung Electronics Co., Ltd.
South Korea

Manufacturer/Client **S**

Smartphone Gamepad Wireless game controller
Category: computer

Manufacturer/Client
Samsung Electronics Co., Ltd.
South Korea

Design
Samsung Electronics Co., Ltd.
South Korea

Vase Standing printer
Category: computer

Manufacturer/Client
Samsung Electronics Co., Ltd.
South Korea

Design
Samsung Electronics Co., Ltd.
South Korea

Egg-tray Supplies packaging
Category: consumer electronics

Manufacturer/Client
Samsung Electronics Co., Ltd.
South Korea

Design
Samsung Electronics Co., Ltd.
South Korea

Samsung Mobile Package Package
Category: consumer electronics

Manufacturer/Client
Samsung Electronics Co., Ltd.
South Korea

Design
Samsung Electronics Co., Ltd.
South Korea

Wireless Audio-Multiroom Network audio system
Category: consumer electronics

Manufacturer/Client
Samsung Electronics Co., Ltd.
South Korea

Design
Samsung Electronics Co., Ltd.
South Korea

PAR 38 lamp LED lamp
Category: lighting

Manufacturer/Client
Samsung Electronics Co., Ltd.
South Korea

Design
Samsung Electronics Co., Ltd.
South Korea

Manufacturer/Client **S**

3050 Series / RB29FE — BMF refrigerator
Category: kitchen

Manufacturer/Client
Samsung Electronics Co., Ltd.
South Korea

Design
Samsung Electronics Co., Ltd.
South Korea

3050 Series / RT38FE — TMF refrigerator
Category: kitchen

Manufacturer/Client
Samsung Electronics Co., Ltd.
South Korea

Design
Samsung Electronics Co., Ltd.
South Korea

FS9000 (FSR) — Refrigerator
Category: kitchen

Manufacturer/Client
Samsung Electronics Co., Ltd.
South Korea

Design
Samsung Electronics Co., Ltd.
South Korea

AX-F500 (AX037FCVAUWD) — Air purifier
Category: household / tableware

Manufacturer/Client
Samsung Electronics Co., Ltd.
South Korea

Design
Samsung Electronics Co., Ltd.
South Korea

VC-F300G Series — Vacuum cleaner
Category: household / tableware

Manufacturer/Client
Samsung Electronics Co., Ltd.
South Korea

Design
Samsung Electronics Co., Ltd.
South Korea

VC-F800G Series — Vacuum cleaner
Category: household / tableware

Manufacturer/Client
Samsung Electronics Co., Ltd.
South Korea

Design
Samsung Electronics Co., Ltd.
South Korea

Manufacturer/Client **S**

LIVE HOLO   Hologram concert
Category: crossmedia – events

Manufacturer/Client
Samsung Everland
South Korea

Design
d'strict
South Korea

SHN-8810   Home network system
Category: buildings

Manufacturer/Client
Samsung SDS Co., Ltd.
South Korea

Design
Samsung SDS Co., Ltd.
South Korea

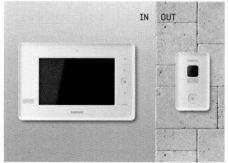

SANTOX S4000-System   Multidesign packaging case
Category: industry / skilled trades

Manufacturer/Client
SANTOX
Germany

Design
SANTOX
Germany

Knowledge Forest   Library system
Category: research+development /professional concepts

Manufacturer/Client
Samsung SDS
South Korea

Design
Samsung SDS
South Korea

ARM-S@NAV   Facade with ventilation
Category: buildings

Manufacturer/Client
Sankyo-Alumi Company
Japan

Design
Plants Associates Inc.
Japan
Sankyo-Alumi Company, Sankyo Tateyama Inc.
Japan

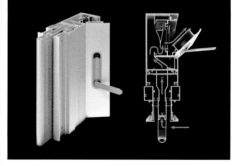

KDDI PortableCharger   Portable charger
Category: telecommunications

Manufacturer/Client
SANYO Electric Co., Ltd.
Japan

Design
SANYO Electric Co., Ltd.
Japan

Manufacturer/Client **S**

KDDI WC01   Wireless charger
Category: telecommunications

Manufacturer/Client
SANYO Electric Co., Ltd.
Japan

Design
SANYO Electric Co., Ltd.
Japan

Universal   Washbasin
Category: bathroom / wellness

Manufacturer/Client
Saturnbath. Co., Ltd.
South Korea

Design
20PLUS
South Korea

DOT Marc Jacobs   Fragrance bottle design
Category: beauty / health / household

Manufacturer/Client
Sayuri Studio, Inc.
Japan

Design
Sayuri Studio, Inc.
Japan

TOCCATA   LED pendant
Category: lighting

Manufacturer/Client
Sattler GmbH
Germany

Design
MARKUS BISCHOF produktdesign
Germany

Varioline Firetube   Sauna cabin
Category: bathroom / wellness

Manufacturer/Client
Saunalux GmbH Products & Co. KG
Germany

Design
bittermann industriedesign
Germany

FN 9280   Compact power entry module
Category: industry / skilled trades

Manufacturer/Client
Schaffner EMV AG
Switzerland

Design
Plast Competence Center AG
Switzerland

Manufacturer/Client **S**

SchauspielhausBochum Website
Category: digital media – public service websites

Manufacturer/Client
Schauspielhaus Bochum
Germany

Design
Oktober
Kommunikationsdesign GmbH
Germany

SCHMIDHUBER Website
Category: digital media – corporate websites

Manufacturer/Client
SCHMIDHUBER
Germany

Design
Hauser Lacour
Germany

Schindelhauer ThinBike Compact urban bicycle
Category: leisure / lifestyle

Manufacturer/Client
Schindelhauer Bikes
Germany

Design
Schindelhauer Bikes
Germany

GOLD

JURY STATEMENT

"This urban bike tackles the concept of urban cycling. It is not a foldable bike but the whole inventive mechanism, which allows you to hang it on the wall, is very smart in a beautiful static way. The bike itself and the use of it are incredibly comfortable and very refined – it is quite monochromatic and very graphic in the execution. A beautiful appearance and a nice aesthetic for a functional object would make you would happy to hang it on the wall."

Miluz Modular switches and sockets
Category: buildings

Manufacturer/Client
Schneider Electric
Brazil

Design
Questto|Nó
Brazil

Zeit statt Zeug Microsite
Category: digital media – microsites

Manufacturer/Client
Scholz & Volkmer GmbH
Germany

Design
Scholz & Volkmer GmbH
Germany

Manufacturer/Client **S**

Allwettermatte F15 All-weather floor mat
Category: transportation design / special vehicles

Manufacturer/Client
Schönek GmbH & Co. KG
Germany

Design
BMW Group Designsworks USA
Germany

C3 Pro Motorcycle helmet
Category: transportation design / special vehicles

Manufacturer/Client
Schuberth GmbH
Germany

Design
Schuberth GmbH
Germany

Schüco: BAU 2013 Trade-fair architecture
Category: corporate architecture – exhibition / trade fair

Manufacturer/Client
Schüco International KG
Germany

Design
D'ART DESIGN GRUPPE GmbH
Germany

AvanTec SimplySmart Fitting system for windows
Category: buildings

Manufacturer/Client
Schüco International KG
Germany

Design
Schüco International KG
Germany

Schüco LightSkin LED lighting
Category: buildings

Manufacturer/Client
Schüco International KG
Germany

Design
Schüco International KG
Germany

Folding Book Book
Category: print media – publishing

Manufacturer/Client
SCIENCE INT'L CO., Ltd.
Taiwan

Design
YUMAMAN
CREATIVE & DESIGN CO., Ltd.
Taiwan

Manufacturer/Client **S**

seepex SCT Progressive cavity pump
Category: industry / skilled trades

Manufacturer/Client
seepex GmbH
Germany

Design
seepex GmbH
Germany

LW-600P Label printer
Category: office / business

Manufacturer/Client
Seiko Epson Corporation
Japan

Design
Seiko Epson Corporation
Japan

Ipiranga Seeds Branding project
Category: crossmedia – corporate design

Manufacturer/Client
Sementes Ipiranga
Brazil

Design
saad branding+design
Brazil

LW-1000P Label printer
Category: office / business

Manufacturer/Client
Seiko Epson Corporation
Japan

Design
Seiko Epson Corporation
Japan

BAMBO LED showcase lighting
Category: lighting

Manufacturer/Client
SELF Electronics Co., Ltd.
China

Design
SELF Electronics Co., Ltd.
China

Jang Sauce packaging
Category: food

Manufacturer/Client
Sempio Foods Company
South Korea

Design
Sempio Foods Company
South Korea

Manufacturer/Client **S**

Century SC 660 / 630   Headset
Category: telecommunications

Manufacturer/Client
Sennheiser Communications A/S
Denmark

Design
BRANDIS Industrial Design
Germany

SP 10 (ML) / SP 20 (ML)   Mobile UC speakerphone
Category: telecommunications

Manufacturer/Client
Sennheiser Communications A/S
Denmark

Design
BRANDIS Industrial Design
Germany

LSP 500 PRO   Integrated PA system
Category: audio / video

Manufacturer/Client
Sennheiser electronic
GmbH & Co. KG
Germany

Design
nr21 DESIGN GmbH
Germany

SENSE   Logo and Visual Identity
Category: print media – corporate design

Manufacturer/Client
Sense
China

Design
Dongdao Design Co., Ltd.
China

SENSE   Packaging design
Category: leisure / lifestyle

Manufacturer/Client
Sense
China

Design
Dongdao Design Co., Ltd.
China

just sauna   Sauna control
Category: bathroom / wellness

Manufacturer/Client
sentiotec GmbH
Austria

Design
design büro groiss peter
Austria

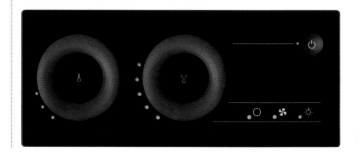

Manufacturer/Client **S**

RC4551   Panorama IP camera
Category: audio / video

Manufacturer/Client
SERCOMM
Taiwan

Design
Sercomm Corporation
Taiwan

Smart Alarm Siren   Alarm system
Category: buildings

Manufacturer/Client
SERCOMM
Taiwan

Design
Sercomm Corporation
Taiwan

Bench Luzia   Bench
Category: living room / bedroom

Manufacturer/Client
Sergio Bertti
Brazil

Design
Ronald Scliar Sasson
Brazil

uDR 580i   Floor-mounted X-ray System
Category: medicine / health+care

Manufacturer/Client
Shanghai United
Imaging Healthcare Co., Ltd.
China

Design
Shanghai United
Imaging Healthcare Co., Ltd.
China

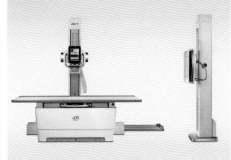

uDR 770i   Ceiling-suspended X-ray system
Category: medicine / health+care

Manufacturer/Client
Shanghai United Imaging Health-
care Co., Ltd.
China

Design
Shanghai United Imaging Health-
care Co., Ltd.
China

FlüssigeTinte   Residential
Category: corporate architecture – architecture / interior design

Manufacturer/Client
Shang Yih
Interior Design Co., Ltd.
Taiwan

Design
Shang Yih
Interior Design Co., Ltd.
Taiwan

Manufacturer/Client **S**

JUSTIME LUCKY 7   Water drinking faucet
Category: kitchen

Manufacturer/Client
Shengtai Brassware Co., Ltd.
Taiwan

Design
JUSTIME Team of Shengtai Brassware
Taiwan

JUSTIME YES   Hose bib
Category: bathroom / wellness

Manufacturer/Client
Shengtai Brassware Co., Ltd.
Taiwan

Design
JUSTIME Team of Shengtai Brassware
Taiwan

KHC63   Horizontal machining center
Category: industry / skilled trades

Manufacturer/Client
Shenji Group
Kunming Machine Tool
China

Design
Fantian Brand Management Consultant Co., Ltd.
China

TK6920B   CNC machine tool
Category: industry / skilled trades

Manufacturer/Client
Shenji Group
Kunming Machine Tool
China

Design
Fantian Brand Management Consultant Co., Ltd.
China

MIPOW BOOMAX   Bluetooth speaker
Category: audio / video

Manufacturer/Client
Shenzhen Baojia Battery Tech Co., Ltd.
China

Design
Shenzhen Baojia Battery Tech Co., Ltd.
China

MIPOW BOOMIN   Bluetooth speaker
Category: audio / video

Manufacturer/Client
Shenzhen Baojia Battery Tech Co., Ltd.
China

Design
Shenzhen Baojia Battery Tech Co., Ltd.
China

Manufacturer/Client **S**

MIPOW BTX200   Bluetooth headphone
Category: audio / video

Manufacturer/Client
Shenzhen Baojia
Battery Tech Co., Ltd.
China

Design
Shenzhen Baojia
Battery Tech Co., Ltd.
China

MIPOW BTX500   Bluetooth headphone
Category: audio / video

Manufacturer/Client
Shenzhen Baojia
Battery Tech Co., Ltd.
China

Design
Shenzhen Baojia
Battery Tech Co., Ltd.
China

VoxTube 700   Bluetooth headset
Category: audio / video

Manufacturer/Client
Shenzhen Baojia
Battery Tech Co., Ltd.
China

Design
Shenzhen Baojia
Battery Tech Co., Ltd.
China

MIPOW Mirror Power   Universal charger
Category: telecommunications

Manufacturer/Client
Shenzhen Baojia
Battery Tech Co., Ltd.
China

Design
Shenzhen Baojia
Battery Tech Co., Ltd.
China

MIPOW SPL05   Universal charger
Category: telecommunications

Manufacturer/Client
Shenzhen Baojia
Battery Tech Co., Ltd.
China

Design
Shenzhen Baojia
Battery Tech Co., Ltd.
China

SPAC03   Wall charger
Category: telecommunications

Manufacturer/Client
Shenzhen Baojia
Battery Tech Co., Ltd.
China

Design
Shenzhen Baojia
Battery Tech Co., Ltd.
China

Manufacturer/Client **S**

iSee4   Eye massager
Category: medicine / health+care

Manufacturer/Client
Shenzhen Breo Technology Co., Ltd.
China

Design
Cube Design China
China
Shenzhen Breo Technology Co., Ltd.
China

PM1218   Desktop air purifier
Category: household / tableware

Manufacturer/Client
Shenzhen Breo Technology Co., Ltd.
China

Design
Shenzhen Jiangyi
Science & Technology
Development Co., Ltd.
China

QVOD Living PC (R810S)   Consumer electronics
Category: audio / video

Manufacturer/Client
Shenzhen QVOD
Technology Co., Ltd.
China

Design
Shenzhen ARTOP Design Co., Ltd.
China

A600   NFC-enabled Bluetooth speaker
Category: audio / video

Manufacturer/Client
Shenzhen Rapoo
Technology Co., Ltd.
China

Design
Shenzhen Rapoo
Technology Co., Ltd.
China

Motion DV   Motion type DV
Category: audio / video

Manufacturer/Client
Shenzhen Shenghongxing
Technology Co., Ltd.
China

Design
Shenzhen Daidea industrial
product design Co., Ltd.
China

Q5 Flashlight Mobile   Mobile phone
Category: telecommunications

Manufacturer/Client
Shenzhen Uoshon Communication
Technology Co., Ltd.
China

Design
Shenzhen Uoshon Communication
Technology Co., Ltd.
China

Manufacturer/Client **S**

DURA-ACE 9000 / 9070
Road-racing bike components, series
Category: leisure / lifestyle

Manufacturer/Client
Shimano Inc.
Japan

Design
Shimano Inc.
Japan

METANIUM Baitcasting reels
Category: leisure / lifestyle

Manufacturer/Client
Shimano Inc.
Japan

Design
Shimano Inc.
Japan

SHIMANO SH-MT44 Touring cycling shoe
Category: leisure / lifestyle

Manufacturer/Client
Shimano Inc.
Japan

Design
Shimano Inc.
Japan

Shimano SLX M670 Mountain bike components, series
Category: leisure / lifestyle

Manufacturer/Client
Shimano Inc.
Japan

Design
Shimano Inc.
Japan

C-Cross Trimaran Cross trimaran
Category: research+development / professional concepts

Manufacturer/Client
Ship and Ocean Industries
R & D Center
Taiwan

Design
Buffon Wang
Taiwan

JETFOIL BUS Vehicle
Category: research+development / professional concepts

Manufacturer/Client
Ship and Ocean Industries
R & D Center
Taiwan

Design
G-WISE DESIGN Co., Ltd.
Taiwan

Manufacturer/Client **S**

UCMB-ZP1 Mobile battery, 3,000 mA 1.5A
Category: telecommunications

Manufacturer/Client
SiB Co., Ltd.
Japan

Design
HYT Design Co., Ltd.
Japan

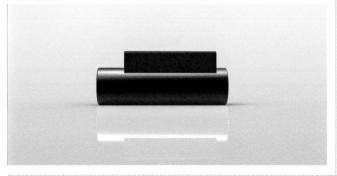

FLOWSIC500 Ultrasonic gas meter
Category: industry / skilled trades

Manufacturer/Client
SICK Engineering GmbH
Germany

Design
code2design
Germany

Alu-Innenausstattung Aluminum interior system
Category: kitchen

Manufacturer/Client
SieMatic Möbelwerke
Germany

Design
speziell®
Germany

Studie zur Energiewende Study
Category: print media – information media

Manufacturer/Client
Siemens AG
Germany

Design
hw.design GmbH
Germany

Cios Alpha Mobile C-arm
Category: medicine / health+care

Manufacturer/Client
Siemens AG
Germany

Design
REFORM DESIGN Produkt
Germany
Siemens AG
Germany

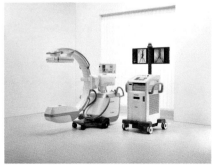

SCALANCE XM400+Extensions
Industrial ethernet switches
Category: industry / skilled trades

Manufacturer/Client
Siemens AG, Industry Sector
Germany

Design
at-design GbR
Germany
Siemens AG, Industry Sector
Germany

Manufacturer/Client **S**

SIMATIC ET200 ECOPN   Automation system, decentral
Category: industry / skilled trades

Manufacturer/Client
Siemens AG, Industry Sector
Germany

Design
at-design GbR
Germany
Siemens AG, Industry Sector
Germany

SIMOTICS FD   Flexible duty motors
Category: industry / skilled trades

Manufacturer/Client
Siemens AG, Industry Sector
Germany

Design
at-design GbR
Germany
Siemens AG, Industry Sector
Germany

SINAMICS G110M   Motor integrated converter
Category: industry / skilled trades

Manufacturer/Client
Siemens AG, Industry Sector
Germany

Design
at-design GbR
Germany
Siemens AG, Industry Sector
Germany

Charismo 2C   Hearing aid
Category: medicine / health+care

Manufacturer/Client
Siemens
Audiologische Technik GmbH
Germany

Design
BRANDIS Industrial Design
Germany

HG73G / HQ/HY738356M   Cooker, line
Category: kitchen

Manufacturer/Client
Siemens Electrogeräte GmbH
Germany

Design
Siemens Electrogeräte GmbH
Germany

KS36FPW30/KS36FPI30   Refrigerator, vitaFresh
Category: kitchen

Manufacturer/Client
Siemens Electrogeräte GmbH
Germany

Design
Siemens Electrogeräte GmbH
Germany

Manufacturer/Client **S**

LC56 / LC86 / LC98 KA Inclined hood, series
Category: kitchen

Manufacturer/Client
Siemens Electrogeräte GmbH
Germany

Design
Siemens Electrogeräte GmbH
Germany

VSQ8 Serie Vacuum cleaner
Category: household / tableware

Manufacturer/Client
Siemens Electrogeräte GmbH
Germany

Design
Siemens Electrogeräte GmbH
Germany

combidome Beverage carton pack
Category: beverages

Manufacturer/Client
SIG Combibloc
Germany

Design
SIG Combibloc
Germany

CrossRoad|TheTurningPoint Commercial space
Category: corporate architecture – hotel / spa / gastronomy

Manufacturer/Client
Sign Architecture & Interior
Design Co., Ltd.
Taiwan

Design
Sign Architecture & Interior
Design Co., Ltd.
Taiwan

Touch 825 USB Flash drive
Category: computer

Manufacturer/Client
Silicon Power
Computer & Communications Inc.
Taiwan

Design
Silicon Power
Computer & Communications Inc.
Taiwan

Laufen Curvetronic Electronic faucet
Category: bathroom / wellness

Manufacturer/Client
Similor AG
Switzerland

Design
platinumdesign
Germany

Manufacturer/Client **S**

adidas – New Walk of Fame   Brand exhibition
Category: corporate architecture – exhibition / trade fair

Manufacturer/Client
simple GmbH
Germany

Design
simple GmbH
Germany

Singapore Airlines Cabin   First class seat and cabin
Category: transportation design / special vehicles

Manufacturer/Client
Singapore Airlines
Singapore

Design
BMW Group DesignworksUSA
United States of America

inEos X5   Extraoral scanner
Category: medicine / health+care

Manufacturer/Client
Sirona Dental Systems GmbH
Germany

Design
Puls Produktdesign
Germany

SK B BOX   Set-Top Box
Category: audio / video

Manufacturer/Client
SK
South Korea

Design
PlusX
South Korea

elements   Chimney stove system
Category: living room / bedroom

Manufacturer/Client
Skantherm Wagner GmbH & Co. KG
Germany

Design
Prof. Wulf Schneider und Partner
Germany

SKF Tool cases   Case for maintenance tools
Category: industry / skilled trades

Manufacturer/Client
SKF Maintenance Products
Netherlands

Design
FLEX/the INNOVATIONLAB B. V.
Netherlands

Manufacturer/Client **S**

SK Planet Brochure Brochure
Category: print media – corporate design

Manufacturer/Client
SK Planet
South Korea

Design
DesignSOHO
South Korea

AIRMENIUS Floorpump
Category: leisure / lifestyle

Manufacturer/Client
SKS metaplast Scheffer-Klute GmbH
Germany

Design
npk design B. V.
Netherlands

OFFSHORE MASTER Safety harness
Category: industry / skilled trades

Manufacturer/Client
SKYLOTEC GmbH
Germany

Design
SKYLOTEC GmbH
Germany

Bicycle folding handlebar Bicycle accessories
Category: leisure / lifestyle

Manufacturer/Client
Smaller International Co., Ltd.
Taiwan

Design
Smaller International Co., Ltd.
Taiwan

EXO Identity design
Category: print media – corporate design

Manufacturer/Client
S.M. Entertainment
South Korea

Design
S.M. Entertainment
South Korea

EXO – XOXO, Growl Album
Category: leisure / lifestyle

Manufacturer/Client
S.M. Entertainment
South Korea

Design
S.M. Entertainment
South Korea

Manufacturer/Client **S**

Girls' Generation Album
Category: leisure / lifestyle

Manufacturer/Client
S.M. Entertainment
South Korea

Design
S.M. Entertainment
South Korea

Girls' Generation Album
Category: leisure / lifestyle

Manufacturer/Client
S.M. Entertainment
South Korea

Design
S.M. Entertainment
South Korea

FOREFRONT All mountain bike helmet
Category: leisure / lifestyle

Manufacturer/Client
Smith Optics
United States of America

Design
Smith Optics
United States of America
Koroyd
Monaco

Isochet Screwdriver
Category: industry / skilled trades

Manufacturer/Client
SNA Europe [Sweden] AB
Sweden

Design
SNA Europe [Sweden] AB
Sweden

Quick Adjust Water pump pliers
Category: industry / skilled trades

Manufacturer/Client
SNA Europe [Sweden] AB
Sweden

Design
SNA Europe [Sweden] AB
Sweden

Appenzeller Exhibition design
Category: corporate architecture – installations in public spaces

Manufacturer/Client
SO Appenzeller Käse GmbH
Switzerland
Appenzeller Schaukäserei
Switzerland
Appenzeller Volkskunde-Museum
Switzerland

Design
Büro4 AG
Switzerland
Curious About
Switzerland

Manufacturer/Client **S**

SodaStream SOURCE Soda carbonizer
Category: household / tableware

Manufacturer/Client
SodaStream
United States of America

Design
Activision
United States of America

Crystal Premium Weiß Drinksmaker
Category: kitchen

Manufacturer/Client
Sodastream Ltd.
Israel

Design
Sodastream Ltd.
Israel

SDL Atrium Carré Glass house
Category: buildings

Manufacturer/Client
SOLARLUX
Aluminium Systeme GmbH
Germany

Design
SOLARLUX
Aluminium Systeme GmbH
Germany

Easy Chair
Category: living room / bedroom

Manufacturer/Client
Sollosbrasil
Brazil

Design
JaderAlmeida design & arquitetura
Brazil

Mad Armchair
Category: living room / bedroom

Manufacturer/Client
Sollosbrasil
Brazil

Design
JaderAlmeida design & arquitetura
Brazil

Soma Water filter with glass carafe
Category: household / tableware

Manufacturer/Client
Soma
United States of America

Design
Soma
United States of America

Manufacturer/Client **S**

SOMA FLEX OPTIMA   Flexographic printing press
Category: industry / skilled trades

Manufacturer/Client	Design
SOMA spol. s. r. o.	IMBUS design
Czech Republic	Czech Republic
	Kokes Partners
	Czech Republic

Sonos PLAY:1   Wireless Hi-Fi system
Category: audio / video

Manufacturer/Client	Design
Sonos Europe B. V.	Sonos Inc.
Netherlands	United States of America

S9 HCU   Color doppler
Category: medicine / health+care

Manufacturer/Client	Design
SonoScape Co., Ltd.	SonoScape Co., Ltd.
China	China

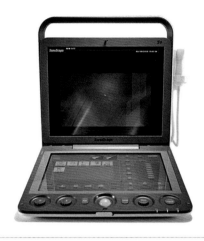

Sonos PLAYBAR™   Soundbar / Wireless Hi-Fi System
Category: audio / video

Manufacturer/Client	Design
Sonos Europe B. V.	Sonos Inc.
Netherlands	United States of America

BRAVIA™ X9000A Series   4K LED HD TV
Category: audio / video

Manufacturer/Client	Design
SONY Corporation	SONY Corporation
Japan	Japan

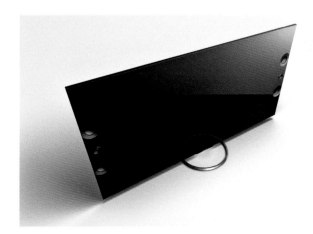

GOLD

JURY STATEMENT

"In this case, we are dealing with a perfect formal reduction of the product type: flat-screen TV. It is aesthetically strong, on the one hand, with its generous format and integrated speakers, yet nonetheless reserved. There are hardly any noticeable details when it is turned off, once switched on, however, it unfolds its overall effect. Every edge, every surface, is perfectly designed and manufactured."

Manufacturer/Client **S**

DSC-RX10   Digital still camera
Category: audio / video

Manufacturer/Client
SONY Corporation
Japan

Design
SONY Corporation
Japan

GOLD

JURY STATEMENT

"The Sony Digital RX10 is the congenial translation of the classical single lens reflex camera form into the year 2014. The casing is unbelievably smooth, and the handling of the camera is virtually self-explanatory. Useless functions are not to be found. The product convinces with strict, strong shapes and an extremely clean design in every detail."

VAIO® Tap11   Tablet
Category: computer

Manufacturer/Client
SONY Corporation
Japan

Design
SONY Corporation
Japan

GOLD

JURY STATEMENT

"This tablet PC convinced us because it reflects a mature brand continuing to innovate with a consistent and coherent design language and with good use of materials, excellent attention to details and overall simplicity and cleanness of execution."

BRAVIA™ W950A Series   LED HD TV
Category: audio / video

Manufacturer/Client
SONY Corporation
Japan

Design
SONY Corporation
Japan

CMT-BT80W / BT60 Series   Home audio system
Category: audio / video

Manufacturer/Client
SONY Corporation
Japan

Design
SONY Europe Ltd.
United Kingdom

Manufacturer/Client **S**

DSC-HX50 Digital still camera
Category: audio / video

Manufacturer/Client
SONY Corporation
Japan

Design
SONY Europe Ltd.
United Kingdom
SONY Corporation
Japan

DSC-QX10 / QX100 Digital still camera
Category: audio / video

Manufacturer/Client
SONY Corporation
Japan

Design
SONY Corporation
Japan

HDR-AS30V w/accessories Digital video camera
Category: audio / video

Manufacturer/Client
SONY Corporation
Japan

Design
SONY Corporation
Japan

HDR-MV1 Music video recorder
Category: audio / video

Manufacturer/Client
SONY Corporation
Japan

Design
SONY Corporation
Japan

HMZ-T3W / HMZ-T3 3D head mounted display
Category: audio / video

Manufacturer/Client
SONY Corporation
Japan

Design
SONY Corporation
Japan

ILCE-7 / ILCE-7R Digital still camera
Category: audio / video

Manufacturer/Client
SONY Corporation
Japan

Design
SONY Corporation
Japan

Manufacturer/Client **S**

MDR-10R Headphones
Category: audio / video

Manufacturer/Client
SONY Corporation
Japan

Design
SONY Europe Ltd.
United Kingdom
SONY Corporation, Japan

NW-ZX1 Digital Media Player
Category: audio / video

Manufacturer/Client
SONY Corporation
Japan

Design
SONY Corporation
Japan

PMW-F55 Digital cinema camera
Category: audio / video

Manufacturer/Client
SONY Corporation
Japan

Design
SONY Corporation
Japan

SRS-BTS50 Personal audio system
Category: audio / video

Manufacturer/Client
SONY Corporation
Japan

Design
SONY Corporation
Japan

SRS-BTV5 Personal audio system
Category: audio / video

Manufacturer/Client
SONY Corporation
Japan

Design
SONY Corporation
Japan

VAIO® Fit 13A PC
Category: computer

Manufacturer/Client
SONY Corporation
Japan

Design
SONY Corporation
Japan

Manufacturer/Client **S**

VAIO® Pro PC
Category: computer

Manufacturer/Client
SONY Corporation
Japan

Design
SONY Corporation
Japan

SH800 / SP6800 Flow cytometer
Category: medicine / health+care

Manufacturer/Client
SONY Corporation
Japan

Design
SONY Corporation
Japan

Crystal Aqua Trees Illumination
Category: corporate architecture – installations in public spaces

Manufacturer/Client
Sony Enterprise Co., Ltd.
Japan

Design
Sony PCL Inc., Japan
WOW Inc., Japan
TORAFU ARCHITECTS, Japan
LUFTZUG CO., Ltd., Netherlands

VAIO® Tap21 PC
Category: computer

Manufacturer/Client
SONY Corporation
Japan

Design
SONY Corporation
Japan

SAXES series Production line
Category: industry / skilled trades

Manufacturer/Client
Sony EMCS Corporation
Japan

Design
SONY Corporation
Japan

Bluetooth Handset SBH52 Smart Bluetooth handset
Category: telecommunications

Manufacturer/Client
Sony Mobile Communications Inc.
Japan

Design
Sony Mobile Communications Inc.
Japan

Manufacturer/Client **S**

Smartwatch2 SW2 Watch
Category: telecommunications

Manufacturer/Client
Sony Mobile Communications Inc.
Japan

Design
Sony Mobile Communications Inc.
Japan

Xperia™ C Smartphone
Category: telecommunications

Manufacturer/Client
Sony Mobile Communications Inc.
Japan

Design
Sony Mobile Communications Inc.
Japan

Xperia Z1f/ Z1 Compact Smartphone
Category: telecommunications

Manufacturer/Client
Sony Mobile Communications Inc.
Japan

Design
Sony Mobile Communications Inc.
Japan

Xperia™ Z Ultra Smartphone
Category: telecommunications

Manufacturer/Client
Sony Mobile Communications Inc.
Japan

Design
Sony Mobile Communications Inc.
Japan

Was war wird – gut Photo book and Annual Report
Category: print media – corporate communication

Manufacturer/Client
Sparkassenverband
Niedersachsen
Germany

Design
KONO
Design und Technologie GmbH
Germany

Book/Swissness Anniversary passport
Category: print media – corporate communication

Manufacturer/Client
Sputnik Engineering AG
Switzerland

Design
weiss communication + design
Switzerland

Manufacturer/Client **S**

BVPC 850-0   Deluxe bus video panel
Category: buildings

Manufacturer/Client
S. Siedle & Söhne
Germany

Design
S. Siedle & Söhne
Germany

NEON   Highlighter
Category: office / business

Manufacturer/Client
STABILO International GmbH
Germany

Design
STABILO International GmbH
Germany

SWM Geschäftsbericht 2012   Annual Report „Love"
Category: print media – corporate communication

Manufacturer/Client
Städtische Werke
Magdeburg GmbH
Germany

Design
wirDesign communications AG
Germany

NEON   Highlighter packaging
Category: leisure / lifestyle

Manufacturer/Client
STABILO International GmbH
Germany

Design
STABILO International GmbH
Germany

Robert by Stadler Form   Air washer
Category: household / tableware

Manufacturer/Client
Stadler Form Aktiengesellschaft
Switzerland

Design
Stadler Form Aktiengesellschaft
Switzerland

Rathaus Schorndorf   Town hall
Category: corporate architecture – architecture / interior design

Manufacturer/Client
Stadt Schorndorf
Germany

Design
Ippolito Fleitz Group –
Identity Architects
Germany
Lichtwerke GmbH
Germany

Manufacturer/Client **S**

Stadt Waldkirch — Brand relaunch
Category: print media – corporate design

Manufacturer/Client
Stadt Waldkirch
Germany

Design
identis, design-gruppe
joseph pölzelbauer
Germany

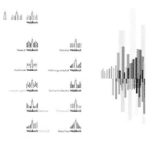

The Pencil — Pencil
Category: office / business

Manufacturer/Client
STAEDTLER Mars GmbH & Co. KG
Germany

Design
STAEDTLER Mars GmbH & Co. KG
Germany

D+D=B — Visual Identity
Category: print media – corporate design

Manufacturer/Client
State Government of Minas Gerais
Brazil

Design
Greco Design
Brazil

GOLD

JURY STATEMENT

"The simplicity of the idea forming the B-logo for the Brazilian design exhibition of two D's – representing diversity and design – has such a recognizable and unique quality that is actually very hard to find. It is a reinvention of typefaces one may never use as they are completely outdated. By combining them with a clean Swiss style, it takes on a new life of its own. The visual quality of the logo is extremely strong while totally simple and pure."

TTS Tool Transport System — Tool trolley
Category: industry / skilled trades

Manufacturer/Client
STAHLWILLE
Germany

Design
STAHLWILLE
Germany

Synfo — Chair for contract segment
Category: office / business

Manufacturer/Client
Stechert Stahlrohrmöbel GmbH
Germany

Design
MARKUS BISCHOF produktdesign
Germany

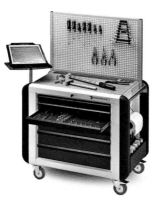

Manufacturer/Client **S**

B-Free Alternative spaces
Category: office / business

Manufacturer/Client
Steelcase S. A.
France

Design
Steelcase S. A.
France

Think Task chair
Category: office / business

Manufacturer/Client
Steelcase S. A.
France

Design
Steelcase S. A.
France

SteelSeries Siberia Elite Gaming headset
Category: audio / video

Manufacturer/Client
SteelSeries ApS
Denmark

Design
SteelSeries ApS
Denmark

Hebensweisheiten Book
Category: print media – publishing

Manufacturer/Client
STEIL KRANARBEITEN
GmbH & Co. KG
Germany

Design
RINGEL & PARTNER
Germany

XSOLAR Solar lights
Category: lighting

Manufacturer/Client
Steinel GmbH
Germany

Design
Kurz Kurz Design
Germany

Emma Tea and coffee series
Category: household / tableware

Manufacturer/Client
Stelton A / S
Denmark

Design
Stelton
Denmark

STIEFELMAYER effective S — Laser cutting machine
Category: industry / skilled trades

Manufacturer/Client
Stiefelmayer-Lasertechnik GmbH & Co
Germany

Design
Saidi 'sign Büro für Produkt- und Grafikdesign
Germany

Museum Luthers Sterbehaus — New permant exhibition
Category: corporate architecture – exhibition / trade fair

Manufacturer/Client
Stiftung Luthergedenkstätten
Germany

Design
neo.studio
Germany

Schlauchreifenverpackung — Tubular packaging
Category: leisure / lifestyle

Manufacturer/Client
STI Group
Germany

Design
STI Group
Germany
KONVERDI GmbH
Germany

Aufbewahrungsbox Leica G-Star — Storage box Leica G-Star
Category: consumer electronics

Manufacturer/Client
STI Group
Germany

Design
G-Star Raw C.V.
Netherlands

dieForm — Design / sales exhibition
Category: corporate architecture – shop / showroom

Manufacturer/Client
stilhaus AG
Switzerland

Design
Gessaga Hindermann GmbH
Switzerland

STILL iGo Easy — Automation App
Category: digital media – mobile applications

Manufacturer/Client
STILL GmbH
Germany

Design
melting elements GmbH
Germany

Manufacturer/Client **S**

Aernario Platinum — Racing bike
Category: leisure / lifestyle

Manufacturer/Client
Storck Bicycle GmbH
Germany

Design
piranha grafik
Germany

GOLD

JURY STATEMENT

"We thought that this racing bike was worthy of an iF gold award because of course the tonality of the product itself, graphically and aesthetically, is very consistent. The shape of the frame, the functionality was impressive and not to forget the optimal ratio between stiffness in relation to weight – the bike is incredibly light, incredibly refined. And for that reason along with the extreme comfort we awarded it gold."

Rebel Seven black edition — Mountain bike
Category: leisure / lifestyle

Manufacturer/Client
Storck Bicycle GmbH
Germany

Design
piranha grafik
Germany

Stryker NAV3i™ — Surgical navigation platform
Category: medicine / health+care

Manufacturer/Client
Stryker Leibinger GmbH & Co. KG
Germany

Design
Erdmann Design
Switzerland

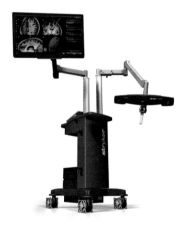

Sudhaus S´elect — Electronic lock
Category: buildings

Manufacturer/Client
Sudhaus GmbH & Co. KG
Germany

Design
Sudhaus GmbH & Co. KG
Germany

Eugens Welt — Browsergame
Category: digital media – online / offline applications

Manufacturer/Client
© Südwestrundfunk
Germany

Design
outermedia GmbH
Germany

Manufacturer/Client **S**

Helium Rigid manual wheelchair
Category: medicine / health+care

Manufacturer/Client
Sunrise Medical GmbH & Co. KG
Germany

Design
Sunrise Medical GmbH & Co. KG
Germany

G•U•M Dental Rinse Foldable bottle
Category: beauty / health / household

Manufacturer/Client
Sunstar Inc.
Japan

Design
IXI Co., Ltd.
Japan

Supermusic Website Redesign
Category: digital media – corporate websites

Manufacturer/Client
[supermusic]
Germany

Design
Falcon White
Germany

Suunto Movescount Online sports community
Category: digital media – corporate websites

Manufacturer/Client
Suunto Oy
Finland

Design
Suunto Design Team and Typolar Oy
Finland

Echo Massage device
Category: leisure / lifestyle

Manufacturer/Client
Svakom Design USA, Limited
United States of America

Design
acme life industrial design Co., Ltd.
China

Gaga Massage device
Category: leisure / lifestyle

Manufacturer/Client
Svakom Design USA, Limited
United States of America

Design
acme life industrial design Co., Ltd.
China

Manufacturer/Client **S**

SWAROSPOT   Downlight
Category: lighting

Manufacturer/Client
Swareflex GmbH
Austria

Design
Swareflex GmbH
Austria

Gold IQnavigator   Handheld control unit
Category: buildings

Manufacturer/Client
Swegon AB
Sweden

Design
Veryday
Sweden

sygonix-Serie   Multiple strips
Category: office / business

Manufacturer/Client
sygonix GmbH
Germany

Design
Lessmandesign
Germany

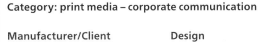

moments   Jubilee book
Category: print media – corporate communication

Manufacturer/Client
Symrise AG
Germany

Design
3st kommunikation GmbH
Germany

SYZYGY Office Frankfurt   Space transforming interior
Category: corporate architecture – architecture / interior design

Manufacturer/Client
SYZYGY Deutschland GmbH
Germany

Design
3deluxe
Germany

tado°   Smart heating control
Category: buildings

Manufacturer/Client
tado° GmbH
Germany

Design
tado° GmbH
Germany

Manufacturer/Client **T**

New clicktap A   Power strip
Category: office / business

Manufacturer/Client
TAEJU Industry. Inc.,
South Korea

Design
we'd design
South Korea

WDC Taipei 2016 Bid Book   Book
Category: print media – information media

Manufacturer/Client
Taiwan Design Center
Taiwan

Design
Freeimage Design
Taiwan

Young Designers' EXPO   Exhibition stand design
Category: crossmedia – advertising / campaigns

Manufacturer/Client
Taiwan Design Center
Taiwan

Design
Redblackdesigns Inc.
Taiwan

Taiwan Noodle House   Restaurant
Category: corporate architecture – architecture / interior design

Manufacturer/Client
Taiwan Noodle House
China

Design
Golucci International Design
Taiwan

Masking Color   Strippable water paint
Category: leisure / lifestyle

Manufacturer/Client
Taiyo Toryo Co., Ltd.
Japan

Design
Ryu Kozeki
Japan

GNZ101N11EX   Premium Nozzle (Handy Type)
Category: leisure / lifestyle

Manufacturer/Client
Takagi Co., Ltd.
Japan

Design
Takagi Co., Ltd.
Japan

Manufacturer/Client **T**

GNZ103N11EX Premium Nozzle (Garden Type)
Category: leisure / lifestyle

Manufacturer/Client
Takagi Co., Ltd.
Japan

Design
Takagi Co., Ltd.
Japan

GNZ104N11 Premium Nozzle (Scrub Type)
Category: leisure / lifestyle

Manufacturer/Client
Takagi Co., Ltd.
Japan

Design
Takagi Co., Ltd.
Japan

State Bathroom cabinet
Category: bathroom / wellness

Manufacturer/Client
talsee
Switzerland

Design
Vetica Group
Switzerland

Tangible Space Corporate Identity
Category: print media – corporate design

Manufacturer/Client
Tangible Space
South Korea

Design
Tangible
South Korea

"Kuromatsu Hakushika" Japanese sake bottle
Category: beverages

Manufacturer/Client
Tatsuuma-Honke Brewing
Company, Ltd.
Japan

Design
Communication Design
Laboratory
Japan

Carus ADAPTO Audio and video indoor stations
Category: buildings

Manufacturer/Client
TCS AG
Germany

Design
TCS AG
Germany
Designhaus p&m
Germany

Manufacturer/Client **T**

Carus IRIS — Audio and video doorstations
Category: buildings

Manufacturer/Client
TCS AG
Germany

Design
TCS AG
Germany
Designhaus p&m
Germany

cubus pure — Living program
Category: living room / bedroom

Manufacturer/Client
TEAM 7 Natürlich Wohnen GmbH
Austria

Design
TEAM 7 Natürlich Wohnen GmbH
Austria

Lightvate® — High-tech safety vest
Category: transportation design / special vehicles

Manufacturer/Client
teamandproducts GmbH
Germany

Design
teamandproducts GmbH
Germany

Tea float — Tea filter
Category: household / tableware

Manufacturer/Client
Teataster Group Holdings Limited
Taiwan

Design
Taiwan

Teatulia Organic Teas — Tea packaging
Category: beverages

Manufacturer/Client
Teatulia Organic Teas
United States of America

Design
Teatulia Organic Teas
United States of America

SVELTE — Portable set-top box
Category: computer

Manufacturer/Client
Technicolor
France

Design
eliumstudio
France

Manufacturer/Client **T**

O2 Live Concept Store Flagship Store
Category: corporate architecture – shop / showroom

Manufacturer/Client
Telefónica Germany
GmbH & Co. OHG
Germany

Design
hartmannvonsiebenthal GmbH
Germany

Tetra Evero® Aseptic Carton bottle
Category: beverages

Manufacturer/Client
Tetra Pak
Italy

Design
Tetra Pak
Italy

Seaside T08 Hydroline Whirlpool bath
Category: bathroom / wellness

Manufacturer/Client
Teuco Guzzini S. p. A.
Italy

Design
Talocci Design
Italy

Tempaline Crowd control barrier
Category: public design

Manufacturer/Client
Tempaline AG, Switzerland
Kunstdünger GmbH, Italy

Design
Plast Competence Center AG
Switzerland

Milestone Duralight® Wash basin
Category: bathroom / wellness

Manufacturer/Client
Teuco Guzzini S. p. A.
Italy

Design
Carlo Colombo Architect
Italy

T-souvenirs Book
Category: print media – publishing

Manufacturer/Client
Teunen Konzepte GmbH
Germany

Design
Fuenfwerken Design AG
Germany

thePrema Presence detector
Category: industry / skilled trades

Manufacturer/Client
Theben AG
Germany

Design
Design Tech
Germany

DOXBOX Lunch box for dogs
Category: research+development / professional concepts

Manufacturer/Client
The Brand Union GmbH
Germany

Design
The Brand Union GmbH
Germany
Werksdesign Volker Schumann
Germany

THIMM Carrycool Insulating shopping box
Category: research+development / professional concepts

Manufacturer/Client
THIMM Verpackung
GmbH & Co. KG
Germany

Design
THIMM Verpackung
GmbH & Co. KG
Germany

Homeless Guerilla campaign
Category: crossmedia – events

Manufacturer/Client
The Brand Union GmbH
Germany

Design
The Brand Union GmbH
Germany

SYBARIS Headset Bluetooth headset
Category: audio / video

Manufacturer/Client
Thermaltake Technology Co., Ltd.
Taiwan

Design
Tt Design
Taiwan

FXD700 / FXD900 Black boxes for vehicles
Category: transportation design / special vehicles

Manufacturer/Client
Thinkware
South Korea

Design
Thinkware
South Korea

Manufacturer/Client **T**

S 1200 Desk
Category: living room / bedroom

Manufacturer/Client
Thonet GmbH
Germany

Design
Thonet GmbH
Germany

THONET Company Website
Category: digital media – corporate websites

Manufacturer/Client
Thonet GmbH
Germany

Design
21TORR GmbH
Germany

Case Logic Luminosity DSLR backpack
Category: leisure / lifestyle

Manufacturer/Client
Thule AB
Sweden

Design
Thule Group
United States of America

Thule RoundTrip Pro Bike transport case
Category: leisure / lifestyle

Manufacturer/Client
Thule Group
Sweden

Design
Thule Group
United States of America

Thüringer Energie Corporate Design
Category: print media – corporate design

Manufacturer/Client
Thüringer Energie AG
Germany

Design
KW43 BRANDDESIGN
Germany

Ideenbuch 2012 Company profile
Category: print media – corporate communication

Manufacturer/Client
ThyssenKrupp Uhde GmbH
Germany

Design
act&react Werbeagentur GmbH
Germany
ThyssenKrupp Uhde GmbH
Germany

Manufacturer/Client T

Tianjin Art Museum   Signage design
Category: corporate architecture – communication media in architecture and public spaces

Manufacturer/Client
Tianjin Art Museum
China

Design
Dongdao Design Co., Ltd.
China

Timesco EES   Laryngoscope
Category: medicine / health+care

Manufacturer/Client
Timesco Ltd.
United Kingdom

Design
PDR
United Kingdom

TRACE + NICE   Track system
Category: lighting

Manufacturer/Client
Tobias Grau GmbH
Germany

Design
Tobias Grau GmbH
Germany

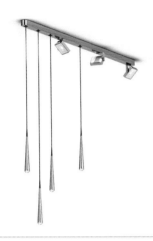

GOLD

JURY STATEMENT

"With this track system, there are multiple choices and each choice brings something special to the environment. One can have directional light and decorative light in the same track. This is a very clean and simple product that adds a unique quality to a room, offering not just light but also decoration. The material quality of the product is noticeably well made and solid."

MONO ergo   Correction tape
Category: office / business

Manufacturer/Client
Tombow Pencil Co., Ltd.
Japan

Design
Tombow Pencil Co., Ltd.
Japan

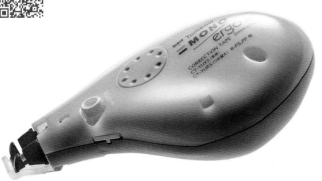

MONO smart   Eraser
Category: office / business

Manufacturer/Client
Tombow Pencil Co., Ltd.
Japan

Design
Tombow Pencil Co., Ltd.
Japan

Manufacturer/Client **T**

Perfect Prep Machine   Baby bottle feed maker
Category: leisure / lifestyle

Manufacturer/Client
Tommee Tippee
United Kingdom

Design
Tommee Tippee
United Kingdom

TomTom 'GO' Series   Personal navigation device
Category: transportation design / special vehicles

Manufacturer/Client
TomTom International B. V.
Netherlands

Design
Native Design
United Kingdom

Runner & Multisport   GPS sports watch
Category: leisure / lifestyle

Manufacturer/Client
TomTom International B. V.
Netherlands

Design
therefore Ltd.
United Kingdom

supraGuide MULTI   Audio and multimedia guide
Category: public design

Manufacturer/Client
tonwelt professional
media GmbH
Germany

Design
EMAMIDESIGN
Germany

Ubike Smart Fold   Folding bike
Category: leisure / lifestyle

Manufacturer/Client
TopGun / Tforce / Ubike
Taiwan

Design
TopGun / Tforce / Ubike
Taiwan

Slash3   QLED TV
Category: audio / video

Manufacturer/Client
Top Victory Electronics
(Taiwan) Co., Ltd.
Taiwan

Design
Top Victory Electronics
(Taiwan) Co., Ltd.
Taiwan

Manufacturer/Client **T**

Tornos SwissNano Swiss type lathe
Category: industry / skilled trades

Manufacturer/Client
Tornos S. A.
Switzerland

Design
Sardi Innovation
Italy

Power TV P2300Series LCD TV
Category: audio / video

Manufacturer/Client
Toshiba Corporation
Japan

Design
Toshiba Corporation
Japan

Qosmio PX30t-A All-in-one computer
Category: computer

Manufacturer/Client
Toshiba Corporation
Japan

Design
Toshiba Corporation
Japan

TCxWave Multi-function retail system
Category: public design

Manufacturer/Client
Toshiba
United States of America

Design
Toshiba
United States of America

REGZA SD-BP1000WP Portable Blu-ray player
Category: audio / video

Manufacturer/Client
Toshiba Corporation
Japan

Design
Toshiba Corporation
Japan

ER-LD530 Convection and steam oven
Category: kitchen

Manufacturer/Client
Toshiba Home
Appliances Corporation
Japan

Design
Toshiba Corporation
Japan

Manufacturer/Client **T**

GR-Y178GD/Y188GD   Refrigerator
Category: kitchen

Manufacturer/Client
Toshiba Home
Appliances Corporation
Japan

Design
Toshiba
Japan

RC-10VWG   IH rice cooker
Category: kitchen

Manufacturer/Client
Toshiba Home
Appliances Corporation
Japan

Design
Toshiba Corporation
Japan

TW-Z96X1   Drum-type washer dryer
Category: household / tableware

Manufacturer/Client
Toshiba Home
Appliances Corporation
Japan

Design
Toshiba Corporation
Japan

VC-S43/VC-S33/VC-S23   Cyclone cleaner
Category: household / tableware

Manufacturer/Client
Toshiba Home
Appliances Corporation
Japan

Design
Toshiba Corporation
Japan

GDR series   Air conditioner
Category: buildings

Manufacturer/Client
Toshiba Home
Appliances Corporation
Japan

Design
Toshiba Corporation
Japan

CONTEMPORARY Faucets   Basin mixer
Category: bathroom / wellness

Manufacturer/Client
TOTO Ltd., Japan
TOTO (China) Co., Ltd., China

Design
TOTO Ltd.
Japan

Manufacturer/Client **T**

C Series Lavatory / Bath   Bathroom series
Category: bathroom / wellness

Manufacturer/Client
TOTO Ltd., Japan
TOTO (China) Co., Ltd., China

Design
TOTO Ltd.
Japan

NEOREST GH / XH / 750H   Shower toilet
Category: bathroom / wellness

Manufacturer/Client
TOTO Ltd., Japan
TOTO (China) Co., Ltd., China
TOTO Europe GmbH, Germany

Design
TOTO Ltd.
Japan

MAGNET BINDER   Magnet stationery
Category: office / business

Manufacturer/Client
TOWA Co., Ltd.
Japan

Design
ATELIER OPA
Japan

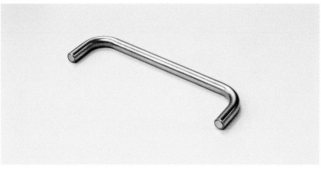

Arlington Visual Budget   Public service website
Category: digital media – public service websites

Manufacturer/Client
Town of Arlington
Financial Office
United States of America

Design
Involution Studios
United States of America

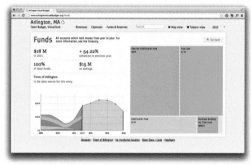

TL-WR706N   Mini WLAN router
Category: computer

Manufacturer/Client
TP-LINK Technologies Co., Ltd.
China

Design
TP-LINK Design
China

sTREEt campaign   Street campaign
Category: print media – corporate design

Manufacturer/Client
TREE PLANET
South Korea

Design
Hancomm, South Korea
INSPIRE.D, South Korea
LIFETHINGS, South Korea
ENSALT, South Korea
Myoung Ho Lee, South Korea
Landon Yoon, South Korea

Manufacturer/Client **T**

sTREEt campaign Campaign
Category: crossmedia – advertising / campaigns

Manufacturer/Client
TREE PLANET
South Korea

Design
Hancomm, South Korea
INSPIRE.D, South Korea
LIFETHINGS, South Korea
ENSALT, South Korea
Myoung Ho Lee, South Korea
Landon Yoon, South Korea

GOLD

JURY STATEMENT

"The concern of this very successful street campaign was to raise awareness about trees in an urban surrounding. The name of the campaign and its very strong logo, which says street and stresses tree within it, is very creative. The placement of the logo is great too because it is always part of the frame, which is the key element of the campaign. As soon as people could see a tree through it, they got involved in the issue and changed their perception of trees."

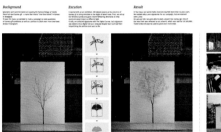

Lateralo Plus LED pendant
Category: lighting

Manufacturer/Client
TRILUX GmbH & Co. KG
Germany

Design
hartmut s. engel design studio
Germany

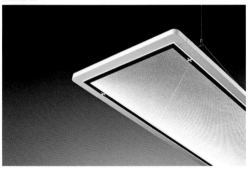

Splitt collection Footwear collection
Category: leisure / lifestyle

Manufacturer/Client
Trippen A. Spieth, M. Oehler GmbH
Germany

Design
Trippen A. Spieth, M. Oehler GmbH
Germany

TRO Corporate Design
Category: print media – corporate design

Manufacturer/Client
TRO GmbH
Germany

Design
KW43 BRANDDESIGN
Germany

GOLD

JURY STATEMENT

"The corporate design of TRO lives from its fantastic bold colour scheme and images, which are so strong that a headline or explanation is not needed. The connection of musician instruments and weapons puts music in the context of power and impact. It is an aesthetic and sensational experience. The images are worked so well that the two objects almost look as one piece. The concept is absolutely consistent and a very self-conscious statement of the company."

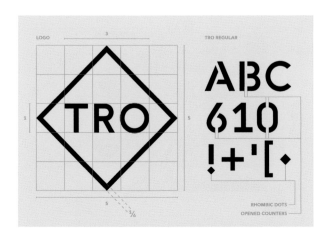

Manufacturer/Client **T**

TRO Posters
Category: print media – advertising media

Manufacturer/Client
TRO GmbH
Germany

Design
KW43 BRANDDESIGN
Germany

true fruits Saft Juice glass carafe
Category: beverages

Manufacturer/Client
true fruits GmbH
Germany

Design
true fruits GmbH
Germany
O-I GLASSPACK GmbH & Co. KG
Germany

TKZP10H13 Cable carrier
Category: industry / skilled trades

Manufacturer/Client
TSUBAKI KABELSCHLEPP GmbH
Germany

Design
TSUBAKIMOTO Chain Company
Japan

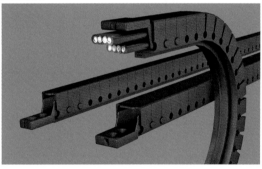

Dream Theater2 Small projector
Category: leisure / lifestyle

Manufacturer/Client
TUNTUN English
South Korea

Design
Neolab convergence
South Korea

Collapsible Cake Taker Adjustable cake taker
Category: household / tableware

Manufacturer/Client
Tupperware France S. A.
France

Design
Tupperware General Services N. V.
Belgium

SmartControl – Panel Operator panel
Category: industry / skilled trades

Manufacturer/Client
Uhlmann Pac-Systeme
GmbH & Co. KG
Germany

Design
CaderaDesign
Germany

Manufacturer/Client **U**

Ultimate Ears BOOM Bluetooth speaker
Category: audio / video

Manufacturer/Client
Ultimate Ears
United States of America

Design
NONOBJECT
United States of America

GOLD

JURY STATEMENT

"The mobile bluetooth speaker has received an iF gold award for a few reasons. One of the main reasons was that it is a consumer product which really looks at the context of the use and the lifestyle of the end user. A very good use of materials, totally water resistant, very durable and of excellent performance and sounds convinces as well."

Safety Case Slim iPhone safety case
Category: telecommunications

Manufacturer/Client
Uncommon LLC
United States of America

Design
GIXIA GROUP Co.
Taiwan

Unikia Knife set
Category: household / tableware

Manufacturer/Client
Unikia S. A.
Norway

Design
fpm
Germany

Moxi Kiss Hearing instrument
Category: medicine / health+care

Manufacturer/Client
Unitron
Canada

Design
Unitron
Canada
AWOL Company
United States of America

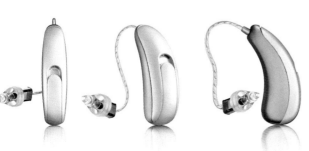

Q&E Ring with Stop Edge Shrink-fit connector
Category: buildings

Manufacturer/Client
Uponor Group
Finland

Design
Uponor Group
Finland

Manufacturer/Client **U**

Anatomie für Kinder   iPad App
Category: digital media – mobile applications

Manufacturer/Client
urbn; interaction
Germany

Design
urbn; interaction
Germany

GOLD

JURY STATEMENT

"This iPad app teaches children all about the human body in a way that is instantly fun and enjoyable. The visual language used in the app tells about human anatomy in a clear and entertaining way. The complex details of the human body have been reduced in such a way that the app feels perfectly customized for the target group."

EasyNAT TB CPA Kit   Molecular diagnostic test set
Category: medicine / health+care

Manufacturer/Client
Ustar Biotechnologies
(Hangzhou) Ltd.
China

Design
Hangzhou Fan design Co., Ltd.
China

HOUR VIEWS   iPad App
Category: digital media – mobile applications

Manufacturer/Client
Vacheron Constantin
Switzerland

Design
KircherBurkhardt GmbH
Germany

aroTHERM   Heat pump
Category: buildings

Manufacturer/Client
Vaillant GmbH
Germany

Design
Vaillant Inhouse Design
Germany

auroCOMPACT   Solarsystem
Category: buildings

Manufacturer/Client
Vaillant GmbH
Germany

Design
Vaillant Inhouse Design
Germany

Manufacturer/Client **V**

ecoCOMPACT Condensing boiler system
Category: buildings

Manufacturer/Client
Vaillant GmbH
Germany

Design
Vaillant Inhouse Design
Germany

Lap®is Electrical socket
Category: material / textiles / wall+floor

Manufacturer/Client
Van den Weghe
Belgium

Design
Van den Weghe
Belgium

vangard Corporate Design
Category: print media – corporate design

Manufacturer/Client
vangard
Germany

Design
labor b designbüro
Germany

TA'OR Organizing system for drawers
Category: living room / bedroom

Manufacturer/Client
Van Hoecke
Belgium

Design
Enthoven Associates
Belgium
Van Hoecke
Belgium

VELUX INTEGRA® System Home automation system
Category: buildings

Manufacturer/Client
VELUX A / S
Denmark

Design
VELUX A / S
Denmark

VEN SPRAY VARIO Spray coating machine
Category: industry / skilled trades

Manufacturer/Client
Venjakob Maschinenbau GmbH
Germany

Design
design Adrian und Greiser GbR
Germany

Manufacturer/Client **V**

Coconuts Concretes Cocos covering tile
Category: material / textiles / wall+floor

Manufacturer/Client
Verdom
Brazil

Design
Verdom
Brazil

Exclusive 6D Refrigerator
Category: kitchen

Manufacturer/Client
Vestel Beyaz Eşya San. ve Tic. A. Ş.
Turkey

Design
VESTEL
Turkey

Viessmann Produktprogramm Heating and energy systems
Category: buildings

Manufacturer/Client
Viessmann Werke GmbH & Co. KG
Germany

Design
Phoenix Design GmbH & Co. KG
Germany

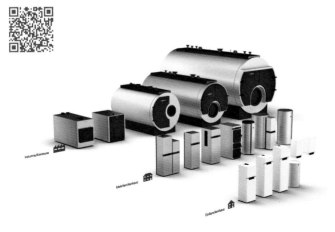

Sprint V Floor cleaning tool
Category: industry / skilled trades

Manufacturer/Client
Vermop Salmon GmbH
Germany

Design
Vermop Salmon GmbH
Germany

PUZZLE Refrigerator
Category: kitchen

Manufacturer/Client
Vestel Beyaz Eşya San. ve Tic. A. Ş.
Turkey

Design
VESTEL
Turkey

Vitomax 300-LT Low pressure hot water boiler
Category: buildings

Manufacturer/Client
Viessmann Werke GmbH & Co. KG
Germany

Design
Phoenix Design GmbH & Co. KG
Germany

Manufacturer/Client **V**

Vitosol 300-T   Vacuum tube collector
Category: buildings

Manufacturer/Client
Viessmann Werke GmbH & Co. KG
Germany

Design
Phoenix Design GmbH & Co. KG
Germany

VIKING MI 632   Robotic lawn mower
Category: leisure / lifestyle

Manufacturer/Client
VIKING GmbH
Austria

Design
Busse Design+Engineering GmbH
Germany

Low Lounger & Sidetable   Lounger with side table
Category: living room / bedroom

Manufacturer/Client
VITEO GmbH
Austria

Design
13&9 Design GmbH
Austria

Copenhagen   Wireless speaker
Category: audio / video

Manufacturer/Client
Vifa Denmark A/S
Denmark

Design
design-people
Denmark

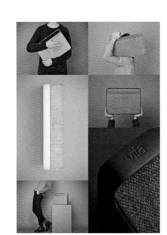

VICLEAN-U   Electronic bidet seat
Category: bathroom / wellness

Manufacturer/Client
Villeroy & Boch AG
Germany

Design
Villeroy & Boch AG
Germany

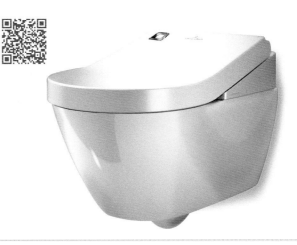

Isotile   External clading tiles
Category: material / textiles / wall+floor

Manufacturer/Client
VitrA Karo
Turkey

Design
VitrA Karo
Turkey

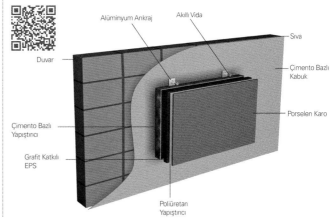

309

Manufacturer/Client **V**

Uptown Porcelain tile
Category: material / textiles / wall+floor

Manufacturer/Client
VitrA Karo
Turkey

Design
VitrA Karo
Turkey

Vichy Ceramic wall tile
Category: material / textiles / wall+floor

Manufacturer/Client
VitrA Karo
Turkey

Design
VitrA Karo
Turkey

Vogel's DesignMount NEXT 7345 wall mount
Category: audio / video

Manufacturer/Client
Vogel's Products B. V.
Netherlands

Design
Vogel's Products B. V.
Netherlands

V-WERKS KATANA Ski
Category: leisure / lifestyle

Manufacturer/Client
Völkl Sports GmbH & Co. KG
Germany

Design
Völkl Sports GmbH & Co. KG
Germany

Volkswagen Golf Variant Passenger car
Category: transportation design / special vehicles

Manufacturer/Client
Volkswagen AG
Germany

Design
Volkswagen AG
Germany

GOLD

JURY STATEMENT

"The very clean and original design of this VW Golf Estate is distinctive and significant for the brand. The design is especially excellent because it does not cry out for excessive emotionality. It rather sets an example of how product design can also achieve a perfect result through simple means. The automobile appears to be, no it is simply of high quality."

Manufacturer/Client **V**

Das Auto. Magazine (Web) Web magazine
Category: digital media – e-zine / e-papers / e-reprots

Manufacturer/Client
Volkswagen AG
Germany

Design
KircherBurkhardt GmbH
Germany

Das Auto. Magazin (App) iPad App
Category: digital media – mobile applications

Manufacturer/Client
Volkswagen AG
Germany

Design
KircherBurkhardt GmbH
Germany

Fleet Magazine App iPad App
Category: digital media – mobile applications

Manufacturer/Client
Volkswagen AG
Germany

Design
Lattke und Lattke GmbH
Germany
Zum Kuckuck GmbH & Co. KG
Germany
COMMANDANTE BERLIN GmbH
Germany

Volkswagen iPad App iPad App
Category: digital media – mobile applications

Manufacturer/Client
Volkswagen AG
Germany

Design
3st kommunikation GmbH
Germany
3st digital GmbH
Germany

Das Auto. Magazin Cross-media magazine
Category: digital media – crossmedia digital

Manufacturer/Client
Volkswagen AG
Germany

Design
KircherBurkhardt GmbH
Germany

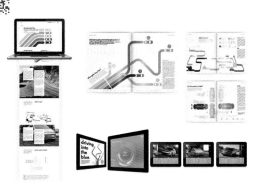

Product Experience 2012 New product launch
Category: crossmedia – events

Manufacturer/Client
Volkswagen AG
Germany

Design
ZIBERT + FRIENDS GmbH
Germany

311

Manufacturer/Client **V**

ECR88D Excavator
Category: transportation design / special vehicles

Manufacturer/Client
Volvo Construction Equipment
South Korea

Design
Volvo Construction Equipment
Sweden

The new Volvo FH Heavy truck
Category: transportation design / special vehicles

Manufacturer/Client
Volvo Trucks
Sweden

Design
Volvo Trucks Product Design
Sweden

VS – Das Klassenzimmer Microsite
Category: digital media – microsites

Manufacturer/Client
VS Vereinigte
Spezialmöbelfabriken
Germany

Design
Zum Kuckuck GmbH & Co. KG
Germany

REFRESH-BUTLER Fabric care
Category: living room / bedroom

Manufacturer/Client
V-ZUG AG
Switzerland

Design
V-ZUG AG
Switzerland

Adora SLQ WP Washing machine
Category: household / tableware

Manufacturer/Client
V-ZUG AG
Switzerland

Design
V-ZUG AG
Switzerland

GOLD

JURY STATEMENT

"The reason that this washing machine is so successful is because it unites several positive qualities: a straightforward dealing with the applied materials in combination with a self-explanatory interface which is extremely easy to operate. Through its clear design language, this washing machine conveys exactly what it is and which functions it has."

Manufacturer/Client **W**

Cintiq Companion Professional creative tablet
Category: computer

Manufacturer/Client
Wacom Company Ltd.
Japan

Design
dingfest I design
Germany

Intuos Pen & Touch Small Tablet for creatives
Category: computer

Manufacturer/Client
Wacom Company Ltd.
Japan

Design
dingfest I design
Germany

Wagner Corporate Website Corporate Website
Category: digital media – corporate websites

Manufacturer/Client
Wagner GmbH
Germany

Design
Martin et Karczinski
Germany

Wagner Architecture Image brochure
Category: print media – corporate design

Manufacturer/Client
Wagner GmbH
Germany

Design
Martin et Karczinski
Germany

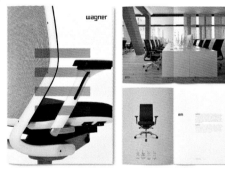

Konkav Design castor
Category: living room / bedroom

Manufacturer/Client
Wagner System GmbH
Germany

Design
Wagner System GmbH
Germany

Screw or Glue Door stopper series
Category: living room / bedroom

Manufacturer/Client
Wagner System GmbH
Germany

Design
Wagner System GmbH
Germany

Manufacturer/Client **W**

Leadchair Executive chair
Category: office / business

Manufacturer/Client
Walter Knoll AG & Co. KG
Germany

Design
EOOS Design GmbH
Austria

K70 Cassette awning
Category: buildings

Manufacturer/Client
WAREMA Renkhoff SE
Germany

Design
WAREMA Renkhoff SE
Germany
CaderaDesign
Germany

Original Magazine
Category: print media – publishing

Manufacturer/Client
Weber-Stephen
Deutschland GmbH
Germany

Design
Fuenfwerken Design AG
Germany

E80 A6 S External venetian blinds
Category: buildings

Manufacturer/Client
WAREMA Renkhoff SE
Germany

Design
WAREMA Renkhoff SE
Germany

Weber Power Control Touch machine control
Category: product interfaces

Manufacturer/Client
Weber Maschinenbau GmbH
Germany

Design
HMI Project
Germany

Viel mehr... Campaign
Category: crossmedia – advertising / campaigns

Manufacturer/Client
Weber-Stephen
Deutschland GmbH
Germany

Design
Fuenfwerken Design AG
Germany

Manufacturer/Client **W**

Weingut Adams Corporate Design
Category: print media – corporate design

Manufacturer/Client
Weingut Adams
Germany

Design
Fuenfwerken Design AG
Germany

Cassita II Cassette awning
Category: buildings

Manufacturer/Client
weinor GmbH & Co. KG
Germany

Design
weinor GmbH & Co. KG
Germany

Opal Design II LED Cassette awning
Category: buildings

Manufacturer/Client
weinor GmbH & Co. KG
Germany

Design
weinor GmbH & Co. KG
Germany

Zenara LED Volant Plus Cassette awning
Category: buildings

Manufacturer/Client
weinor GmbH & Co. KG
Germany

Design
weinor GmbH & Co. KG
Germany

LUMOS Selección Wine packaging
Category: beverages

Manufacturer/Client
Wein & Vinos GmbH
Germany

Design
Ruska, Martín, Associates GmbH
Germany

HCL-301 Heating Airpad Heating pad
Category: leisure / lifestyle

Manufacturer/Client
WEIREI INTERNATIONAL Co., Ltd.
Taiwan

Design
GIXIA GROUP Co.
Taiwan

LABterminal   Laboratory furniture system
Category: medicine / health+care

Manufacturer/Client
Wesemann GmbH
Germany

Design
h&h design GmbH
Germany

INDEFINITE   Residential
Category: corporate architecture – architecture / interior design

Manufacturer/Client
WE & WIN CONSTRUCTION
Taiwan

Design
Wen Sheng Lee
Architects & Planners
Taiwan

gesis FLEX   Room automation
Category: buildings

Manufacturer/Client
Wieland Electric GmbH
Germany

Design
[d]tom
Germany

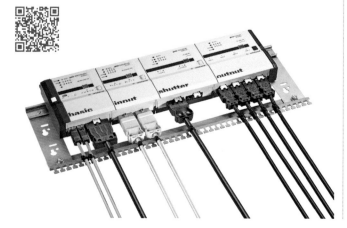

Dimple Cylinder 941   Lock system
Category: buildings

Manufacturer/Client
WEST CORPORATION
Japan

Design
WEST CORPORATION
Japan

Ares Combo   Front-load washer-dryer
Category: household / tableware

Manufacturer/Client
Whirlpool (China)
Investment Co., Ltd.
China

Design
Global Consumer Design,
Whirlpool
China

Magazin-Bithalter Family   Magazine bit holder
Category: industry / skilled trades

Manufacturer/Client
Wiha Werkzeuge GmbH
Germany

Design
Wiha Werkzeuge GmbH
Germany

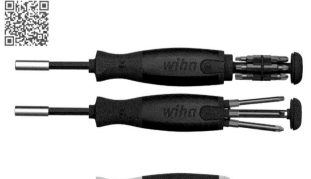

Manufacturer/Client **W**

Worx Transparency Transparent LED pendant
Category: lighting

Manufacturer/Client
WILA Lichttechnik
Germany

Design
WILA Lichttechnik
Germany

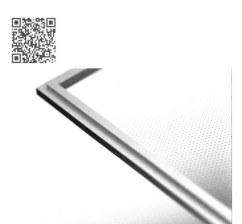

Graph Table
Category: office / business

Manufacturer/Client
Wilkhahn
Germany

Design
jehs + laub GbR
Germany

novum Magazine cover
Category: print media – publishing

Manufacturer/Client
will Magazine Verlag GmbH
Germany

Design
Clormann Design GmbH
Germany

Wilo-Yonos MAXO Glandless circulation pump
Category: buildings

Manufacturer/Client
WILO SE
Germany

Design
Mehnert Corporate Design
GmbH & Co. KG
Germany

Wingman Condom with plastic wings
Category: medicine / health+care

Manufacturer/Client
Wingman Condoms B. V.
Netherlands

Design
Wingman Condoms B. V.
Netherlands
DI – Design Industrial Ltda.
Brazil

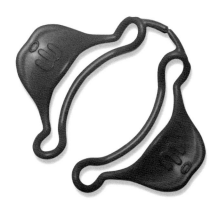

GOLD

JURY STATEMENT

"This condom is an important contribution to our health system. The special design is conducive to a condom being 100 % safe and to it being applied without causing any damage. It is easy to use and due to its analogy to pouting lips, the design is also visually appealing. The design is an innovation in an area where it has a real effect."

Manufacturer/Client **W**

HAIBIKE AFFAIR RX   Carbon racing bike
Category: leisure / lifestyle

Manufacturer/Client
Winora-Staiger GmbH
Germany

Design
Haibike Design Team
Germany

HAIBIKE HEET RX   All mountain bike
Category: leisure / lifestyle

Manufacturer/Client
Winora-Staiger GmbH
Germany

Design
Haibike Design Team
Germany

HAIBIKE XDURO NDURO PRO   Enduro e-mountainbike
Category: leisure / lifestyle

Manufacturer/Client
Winora-Staiger GmbH
Germany

Design
Kucher & Thusbass
Germany
Winora Engineering Team
Germany

HAIBIKE XDURO RACE   E-race bike
Category: leisure / lifestyle

Manufacturer/Client
Winora-Staiger GmbH
Germany

Design
Kucher & Thusbass
Germany
Haibike Design Team
Germany

HAIBIKE XDURO URBAN   Urban pedelec
Category: leisure / lifestyle

Manufacturer/Client
Winora-Staiger GmbH
Germany

Design
Kucher & Thusbass
Germany
Haibike Design Team
Germany

mobile food and foodhunting   Book design
Category: print media – corporate communication

Manufacturer/Client
winwood48edition.com
Germany

Design
identis, design-gruppe joseph pölzelbauer
Germany

Manufacturer/Client **W**

Withings Pulse — Activity tracker
Category: leisure / lifestyle

Manufacturer/Client
Withings
France

Design
EliumStudio
France

OXYGEN Jersey — Roadbike jersey
Category: leisure / lifestyle

Manufacturer/Client
W.L. Gore & Associates GmbH
Germany

Design
W.L. Gore & Associates GmbH
Germany

OXYGEN SO Jersey — Roadbike windstopper jersey
Category: leisure / lifestyle

Manufacturer/Client
W.L. Gore & Associates GmbH
Germany

Design
W.L. Gore & Associates GmbH
Germany

wodtke ixpower — Pellet stove
Category: buildings

Manufacturer/Client
wodtke GmbH
Germany

Design
wodtke GmbH
Germany

Windstärke 10 — Corporate Design
Category: print media – corporate design

Manufacturer/Client
Wrack- und Fischereimuseum
Cuxhaven
Germany

Design
JUNO
Germany

F-50JL — Water heater
Category: buildings

Manufacturer/Client
WUHU MIDEA
KITCHEN AND BATH
China

Design
WUHU MIDEA
KITCHEN AND BATH
China

Manufacturer/Client **W**

HP4S Water heater
Category: buildings

Manufacturer/Client
WUHU MIDEA
KITCHEN AND BATH
China

Design
Feish Design Co., Ltd., China
WUHU MIDEA
KITCHEN AND BATH
China

LE5Q Water heater
Category: buildings

Manufacturer/Client
WUHU MIDEA
KITCHEN AND BATH
China

Design
WUHU MIDEA
KITCHEN AND BATH
China

ASMO read Recessed luminaire
Category: lighting

Manufacturer/Client
XAL GmbH
Austria

Design
Reflexion ag | Thomas Mika
Switzerland

HZ2 Water heater
Category: buildings

Manufacturer/Client
WUHU MIDEA
KITCHEN AND BATH
China

Design
WUHU MIDEA
KITCHEN AND BATH
China

Little Swan TD85-1406VID Front loading washing machine
Category: household / tableware

Manufacturer/Client
WUXI little swan company limited
China

Design
WUXI little swan company limited
China
designaffairs GmbH
Germany

AURO drop Recessed spotlight
Category: lighting

Manufacturer/Client
XAL GmbH
Austria

Design
Reflexion ag | Thomas Mika
Switzerland

Manufacturer/Client **X**

BULLET Track light system
Category: lighting

Manufacturer/Client
XAL GmbH
Austria

Design
XAL GmbH
Austria

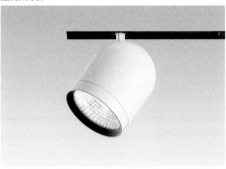

DOC Outdoor luminaire
Category: lighting

Manufacturer/Client
XAL GmbH
Austria

Design
XAL GmbH
Austria

HELIOS suspended Suspended luminaire
Category: lighting

Manufacturer/Client
XAL GmbH
Austria

Design
Kai Stania Product Design
Austria

CIRO ceiling Surface mounted luminaire
Category: lighting

Manufacturer/Client
XAL GmbH
Austria

Design
Kai Stania Product Design
Austria

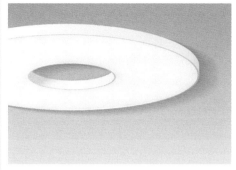

HELIOS floor Floor luminaire
Category: lighting

Manufacturer/Client
XAL GmbH
Austria

Design
Kai Stania Product Design
Austria

Effektor Golf Arm Precizer Hightech arm warmer
Category: leisure / lifestyle

Manufacturer/Client
X-BIONIC®
Italy

Design
X-Technology Swiss R & D AG
Switzerland

Manufacturer/Client **X**

Effektor Power Gloves   Bike and MTB gloves
Category: leisure / lifestyle

Manufacturer/Client
X-BIONIC®
Italy

Design
X-Technology Swiss R & D AG
Switzerland

Energizer Sports Bra   Hightech sports bra
Category: leisure / lifestyle

Manufacturer/Client
X-BIONIC®
Italy

Design
X-Technology Swiss R & D AG
Switzerland

X-BIONIC® Radiactor EVO   Hightech ski underwear
Category: leisure / lifestyle

Manufacturer/Client
X-BIONIC®
Italy

Design
X-Technology Swiss R & D AG
Switzerland

X-BIONIC® The Trick   Hightech running and bike wear
Category: leisure / lifestyle

Manufacturer/Client
X-BIONIC®
Italy

Design
X-Technology Swiss R & D AG
Switzerland

X-BIONIC® Xceed   Hightech golf underwear
Category: leisure / lifestyle

Manufacturer/Client
X-BIONIC®
Italy

Design
X-Technology Swiss R & D AG
Switzerland

Smart Turn Series   Handshower
Category: bathroom / wellness

Manufacturer/Client
Xiamen Solex Technology Co., Ltd.
China

Design
Xiamen Solex Technology Co., Ltd.
China

Manufacturer/Client **X**

Wing Basin faucet
Category: bathroom / wellness

Manufacturer/Client
Xiamen Solex Technology Co., Ltd.
China

Design
Xiamen Solex Technology Co., Ltd.
China

Odin Safety torch
Category: transportation design / special vehicles

Manufacturer/Client
Xindao
Netherlands

Design
Xindao
Netherlands

Thor Safety hammer
Category: transportation design / special vehicles

Manufacturer/Client
Xindao
Netherlands

Design
Xindao
Netherlands

Port Solar charger
Category: telecommunications

Manufacturer/Client
Xindao
Netherlands

Design
Xindao
Netherlands

LIGHTWELL Residential
Category: corporate architecture – architecture / interior design

Manufacturer/Client
Xing Fu Cheng
Taiwan

Design
WEN SHENG LEE ARCHITECTS & PLANNERS
Taiwan

EDO Tableware
Category: household / tableware

Manufacturer/Client
X-TEC GmbH
Austria

Design
X-TEC GmbH
Austria

Manufacturer/Client **Y**

Whispbar WB201   Upright bicycle carrier
Category: transportation design / special vehicles

Manufacturer/Client
Yakima Products, Inc.
United States of America

Design
Yakima Products, Inc.
United States of America

Whispbar WB401   Kayak carrier
Category: transportation design / special vehicles

Manufacturer/Client
Yakima Products, Inc.
United States of America

Design
Yakima Products, Inc.
United States of America

SILENT Brass™   Muting system for brass instruments
Category: leisure / lifestyle

Manufacturer/Client
Yamaha Corporation
Japan

Design
Yamaha Corporation
Japan

CL Series   Digital mixing console
Category: audio / video

Manufacturer/Client
Yamaha Corporation
Japan

Design
Yamaha Corporation
Japan

MT-09   Motorcycle
Category: transportation design / special vehicles

Manufacturer/Client
YAMAHA MOTOR CO., Ltd.
Japan

Design
GK Dynamics Inc.
Japan

YAUCHI   Kitchen knife
Category: household / tableware

Manufacturer/Client
Yauchi Hamono
Japan

Design
ripple effect Co., Ltd.
Japan

Manufacturer/Client **Y**

YG Entertainment — Corporate Identity
Category: print media – corporate design

Manufacturer/Client
YG Entertainment
South Korea

Design
PlusX
South Korea

Yishouyuan — Packaging design
Category: research+development / professional concepts

Manufacturer/Client
Yishouyuan
Bienenprodukte GmbH
China

Design
Dongdao Design Co., Ltd.
China

Y NOT — Retail space design
Category: corporate architecture – shop / showroom

Manufacturer/Client
YNOT – HK (YMO)
Hong Kong

Design
Andy Tong Interiors Ltd.
Hong Kong

Yoga / Tina Schmincke — Business papers
Category: print media – corporate design

Manufacturer/Client
Yoga
Germany

Design
Mit Biss.
Germany

Scorching waste; Casting — Interior Design
Category: corporate architecture – architecture / interior design

Manufacturer/Client
YUN-YIH DESIGN COMPANY
Taiwan

Design
YUN-YIH DESIGN COMPANY
Taiwan

RUNTAL COSMOPOLITAN — Designer radiator
Category: buildings

Manufacturer/Client
Zehnder Group
Boleslawiec Sp. z o. o
Poland

Design
King & Miranda Design S. R. L.
Italy

Manufacturer/Client **Z**

Traue Deinen Sinnen Key visual for MCBW lecture
Category: print media – corporate design

Manufacturer/Client	Design
Zeichen & Wunder GmbH	Zeichen & Wunder GmbH
Germany	Germany

Zeutschel zeta Office scanner
Category: computer

Manufacturer/Client	Design
Zeutschel GmbH	Vetica Group
Germany	Switzerland

Dahua 3X Zoom Camera Full HD network IR camera
Category: audio / video

Manufacturer/Client	Design
Zhejiang Dahua Technology Co., Ltd.	Zhejiang Dahua Technology Co., Ltd.
China	China

Dahua G10 Series Intelligent home system
Category: buildings

Manufacturer/Client	Design
Zhejiang Dahua Technology Co., Ltd.	Zhejiang Dahua Technology Co., Ltd.
China	China

XY-211 Smart blood pressure monitor
Category: medicine / health+care

Manufacturer/Client	Design
Zhejiang Medzone Medical Devices	Institute of Zhejiang Medzone
China	China
	AUG design Co., Ltd.
	China

XY-311 Smart pulse oximeter
Category: medicine / health+care

Manufacturer/Client	Design
Zhejiang Medzone Medical Devices	Institute of Zhejiang Medzone
China	China
	AUG design Co., Ltd.
	China

Manufacturer/Client **Z**

Marshall Monitor Headphones
Category: audio / video

Manufacturer/Client
Zound Industries
Sweden

Design
Zound Industries
Sweden

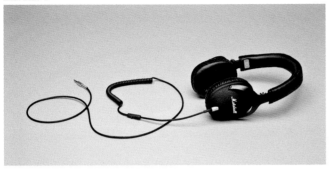

Urbanears Kransen Headphones
Category: audio / video

Manufacturer/Client
Zound Industries
Sweden

Design
Zound Industries
Sweden

Geschäftsbericht 2011/12 Book
Category: print media – corporate communication

Manufacturer/Client
Zumtobel AG
Austria

Design
Anish Kapoor Studio
United Kingdom
Brighten the Corners Ltd.
United Kingdom

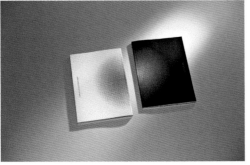

Urbanears Humlan Headphones
Category: audio / video

Manufacturer/Client
Zound Industries
Sweden

Design
Zound Industries
Sweden

ZING Smartphone
Category: telecommunications

Manufacturer/Client
ZTE Corporation
China

Design
ZTE Corporation
China

ARCOS Xpert LED spotlight
Category: lighting

Manufacturer/Client
Zumtobel Lighting GmbH
Austria

Design
David Chipperfield Architects Ltd.
United Kingdom

Manufacturer/Client **Z**

DIAMO LED downlight
Category: lighting

Manufacturer/Client
Zumtobel Lighting GmbH
Austria

Design
Bartenbach GmbH
Austria
Zumtobel Lighting GmbH
Austria

GRAFT High-bay LED luminaire
Category: lighting

Manufacturer/Client
Zumtobel Lighting GmbH
Austria

Design
ARUP
United Kingdom

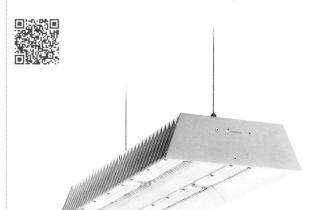

LIGHT FIELDS evolution LED luminaire range
Category: lighting

Manufacturer/Client
Zumtobel Lighting GmbH
Austria

Design
Christopher Redfern
Italy

LINCOR LED pendant and surface luminaire
Category: lighting

Manufacturer/Client
Zumtobel Lighting GmbH
Austria

Design
Zumtobel Lighting GmbH
Austria

LED LENSER® M17R LED torch rechargeable
Category: lighting

Manufacturer/Client
Zweibrüder Optoelectronics
Germany

Design
Zweibrüder Optoelectronics
Germany

LED LENSER P7.2 LED flash light
Category: lighting

Manufacturer/Client
Zweibrüder Optoelectronics
Germany

Design
Zweibrüder Optoelectronics
Germany

Manufacturer/Client **Z**

LED LENSER SEO-Serie LED headlamp series
Category: lighting

Manufacturer/Client
Zweibrüder Optoelectronics
Germany

Design
Zweibrüder Optoelectronics
Germany

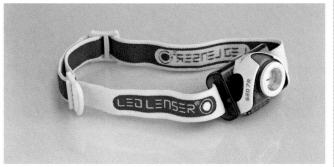

iF INDUSTRIE FORUM DESIGN e.V.

MITGLIEDERLISTE | *LIST OF MEMBERS*

VORSTÄNDE / BOARD
1. VORSITZENDER / 1ST CHAIRMAN

Ernst Raue
ER GBS GmbH
Expo Plaza 11, World Trade Center
30539 Hannover
Germany
www.er-gbs.com

Christoph Böninger
Böcklinstraße 54
80638 München
Germany
www.auerberg.eu

Prof. Fritz Frenkler
f/p design GmbH
Mauerkicherstraße 4
81679 München
Germany
www.fp-design-gmbh.com

Susanne Schmidhuber
SCHMIDHUBER
Nederlinger Straße 21
80638 München
Germany
www.schmidhuber.de

Boris Kochan
KOCHAN & PARTNER GmbH
Strategie | Kommunikation | Design
Hirschgartenallee 25
80639 München
Germany
www.kochan.de

GESCHÄFTSFÜHRENDER VORSTAND / MANAGING DIRECTOR

Ralph Wiegmann
iF Industrie Forum Design e.V.
Bahnhofstrasse 8
30159 Hannover
Germany
www.ifdesign.de

EHRENMITGLIEDER / HONORARY MEMBERS

Herbert H. Schultes
Rodelbahnstraße 1
82256 Fürstenfeldbruck
Germany

Knut Bliesener
Im Haspelfelde 31
30173 Hannover
Germany

Prof. Herbert Lindinger
Brahmstraße 3
30175 Hannover
Germany

Sepp D. Heckmann
Schopenhauerstraße 12
30625 Hannover
Germany

UNTERNEHMEN / CORPORATE MEMBERS

AL-KO Geräte GmbH
Ichenhauser Straße 14
89359 Kötz
Germany
www.al-ko.de

Artemide GmbH
Hans-Böckler-Straße 2
58730 Fröndenberg
Germany
www.artemide.de

B / S / H Bosch und Siemens
Hausgeräte GmbH
Carl-Wery-Straße 34
81793 München
Germany
www.bsh-group.de

BASF SE
Designfabrik E-KTE/IF – F206, Zi. 563
E-KTE/IF – F 206, Zi. 563
67056 Ludwigshafen
Germany
www.basf.com

Blomus GmbH
Zur Hubertushalle 4
59846 Sundern
Germany
www.blomus.com

BMW AG
Knorrstraße 147
80788 München
Germany
www.bmwgroup.com

BOS Best of Steel
Lütkenfelde 4
48282 Emstetten
Germany
www.bestofsteel.de

BREE Collection GmbH & Co. KG
Gerberstraße 3
30916 Isernhagen
Germany
www.bree.com

Carpet Concept
Objekt-Teppichboden GmbH
Bunzlauer Straße 7
33719 Bielefeld
Germany
www.carpet-concept.de

CLAGE GmbH
Pirolweg 1-5
21337 Lüneburg
Germany
www.clage.de

Crown Gabelstapler GmbH & Co. KG
Moosacher Straße 52
80809 München
Germany
www.crown.com/de

Daimler AG
Mercedes-Benz Cars Develpoment
Werk 059, HPC X800
71059 Sindelfingen
Germany
www.daimler.com

designaffairs GmbH
Balanstraße 73
81541 München
Germany
www.designaffairs.com

Deutsche Messe AG
Messegelände
30521 Hannover
Germany
www.messe.de

EnBW AG
Messen und Events
Obere Stegwiesen 29
88400 Biberach
Germany
www.enbw.com

EnBW AG
Marketing und Koordination
Durlacher Allee 93
76131 Karlsruhe
Germany
www.enbw.com

Expotechnik Heinz Soschinski GmbH
Aarstraße 176
65232 Taunusstein
Germany
www.expotechnik.com

FESTO AG & Co. KG
Ruiter Straße 82
73734 Esslingen
Germany
www.festo.com

Gira Giersiepen GmbH & Co. KG
Dahlienstraße
42477 Radevormwald
Germany
www.gira.de

Grohe AG
Design Studio
Feldmühleplatz 15
40545 Düsseldorf
Germany
www.grohe.com

häfelinger + wagner design gmbh
Türkenstraße 55-57
80799 München
Germany
www.hwdesign.de

Hansgrohe SE
Auestraße 5-9
77761 Schiltach
Germany
www.hansgrohe.com

HEWI Heinrich Wilke GmbH
Postfach 1260
34442 Bad Arolsen
Germany
www.hewi.de

Hiller Objektmöbel GmbH & Co. KG
Kippenheimer Straße 6
77971 Kippenheim
Germany
www.hiller-moebel.de

IBM Deutschland GmbH
IBM-Allee 1
71139 Ehningen
Germany
www.ibm.com/de

interstil – Diedrichsen GmbH
Liebigstraße 1-3
33803 Steinhagen
Germany
www.interstil.de

Interstuhl Büromöbel GmbH & Co. KG
Brühlstraße 21
72469 Meßstetten-Tieringen
Germany
www.interstuhl.de

Isaria Corporate Design AG
Gewerbepark Aich 7-9
85667 Oberpframmern
Germany
www.isaria.com

JAB Teppiche Heinz Anstoetz KG
Dammheider Straße 67
32052 Herford
Germany
www.jab.de

Kermi GmbH
Pankofen-Bahnhof 1
94447 Plattling
Germany
www.kermi.de

KOCHAN & PARTNER GmbH
Strategie | Kommunikation | Design
Hirschgartenallee 25
80639 München
Germany
www.kochan.de

Köttermann GmbH & Co. KG
Industriestraße 2-10
31311 Uetze
Germany
www.koettermann.com

KWC AG
Hauptstraße 57
5726 Unterkulm
Switzerland
www.kwc.ch

MüllerKälber GmbH
Daimlerstraße 2
71546 Aspach
Germany
www.muellerkaelber.de

Niedersächsisches Ministerium für
Wirtschaft, Technologie u. Verkehr
Friedrichswall 1
30159 Hannover
Germany
www.mw.niedersachsen.de

Nils Holger Moormann GmbH
An der Festhalle 2
83229 Aschau
Germany
www.moormann.de

OCTANORM-Vertriebs-GmbH
für Bauelemente
Raiffeisenstraße 39
70794 Filderstadt
Germany
www.octanorm.de

Oventrop GmbH & Co. KG
Paul-Oventrop-Straße 1
59939 Olsberg
Germany
www.oventrop.de

Philips GmbH
UB Consumer Lifestyle
Lübeckertordamm 5
20099 Hamburg
Germany
www.philips.de

PLANMECA OY
Asentajankatu 6
00880 Helsinki
Finland
www.planmeca.com

Poggenpohl Möbelwerke GmbH
Poggenpohl Straße 1
32051 Herford
Germany
www.poggenpohl.com

REALTIME TECHNOLOGY (RTT)
Rosenheimer Straße 145
81671 München
Germany
www.rtt.ag

Schneider Electric GmbH
Gothaer Str. 29
40880 Ratingen
Germany
www.schneider-electric.com

SCHULTE Duschkabinenbau KG
Lockweg 81
59846 Sundern
Germany

Sedus Stoll AG
Brückenstraße 15
79761 Waldshut-Tiengen
Germany
www.sedus.de

Sennheiser electronic GmbH & Co. KG
Am Labor 1
30900 Wedemark
Germany
www.sennheiser.com

Siemens AG
Abt. CC MC1
Wittelsbacherplatz 2
80333 München
Germany
www.siemens.com

SKYLOTEC GmbH
Im Bruch 11-15
56567 Neuwied
Germany
www.skylotec.de

SMA Solar Technology AG
Sonnenallee 1
34266 Niestetal
Germany
www.sma.de

Steelcase Werndl AG
Georg-Aicher-Straße 7
83026 Rosenheim
Germany
www.steelcase-werndl.de

Steinberg GmbH
Schiess Str. 30
40549 Düsseldorf
Germany
www.steinberg-armaturen.de

Storck Bicycle GmbH
Rudolfstraße 1
65510 Idstein
Germany
www.storck-bicycle.de

TRILUX GmbH + Co KG
Heidestraße 4
59759 Arnsberg
Germany
www.trilux.de

Vauth-Sagel
Systemtechnik GmbH & Co. KG
Neue Straße 27
33034 Brakel-Erkeln
Germany
www.vauth-sagel.de

Viessmann Werke GmbH Co. KG
Viessmannstraße 1
35108 Allendorf/Eder
Germany
www.viessmann.de

Volkswagen AG
Volkswagen Design
Brieffach 1701, Berliner Ring 2
38436 Wolfsburg
Germany
www.volkswagen.de

wiege Entwicklungs GmbH
Hauptstraße 81
31848 Bad Münder
Germany
www.wiege.com

Wilkhahn
Wilkening + Hahne GmbH + Co. KG.
Fritz Hahne Straße 8
31848 Bad Münder
Germany
www.wilkhahn.de

wodtke GmbH
Rittweg 55-57
72070 Tübingen-Hirschau
Germany
www.wodtke.com

Yokogawa Electric Corporation
2-9-32 naka-cho
Mushashino-shi, Tokyo, 180-8750
Japan
www.yokogawa.com

Zumtobel Lighting GmbH
Schweizer Straße 30
6850 Dornbirn
Austria
www.zumtobel.com

DESIGNER O. DESIGNBÜROS /
DESIGN STUDIOS OR DESIGNER

3-point concepts GmbH
Prinzessinnenstr. 1
10969 Berlin
Germany
www.3pc.de

.molldesign
Turmgasse 7
73525 Schwäbisch Gmünd
Germany
www.molldesign.de

aka buna design consult
Nonnbergstiege 1
5020 Salzburg
Austria
www.akabuna.at

ArGe Design braucht Täter
c/o id 3 Michael Gehlen
Ägidiusstraße 1
50937 Köln
Germany
www.martin-neuhaus.com

artcollin
Goethestraße 28
80336 München
Germany
www.artcollin.de

ArteFakt Pohl Fiegl
Liebigstraße 50 – 52
64293 Darmstadt
Germany
www.artefakt.de

at-design
Flugplatzstraße 111
90768 Fürth
Germany
www.atdesign.de

Avidus Consulting
Wielandstraße 1
65187 Wiesbaden
Germany
www.avidus-consulting.com

bayern design forum e.V.
Luitpoldstraße 3
90402 Nürnberg
Germany
www.bayern-design.de

B:SiGN Design & Communications GmbH
Ellernstraße 36
30175 Hannover
Germany
www.bsign.de

Braake Design
Turnierstraße 3
70599 Stuttgart
Germany
www.braake.com

D´ART Visuelle Kommunikation GmbH
Adlerstraße 41
70199 Stuttgart
Germany
www.dartwork.de

design studio hartmut s. engel
Monreposstraße 7
71634 Ludwigsburg
Germany
www.designstudioengel.de

Design Tech
Zeppelinstraße 53
72119 Ammerbuch
Germany
www.designtech.eu

Deutscher Designer Club e.V.
Große Fischestraße 7
60311 Frankfurt am Main
Germany
www.ddc.de

DRWA Das Rudel Werbeagentur
Erbprinzenstraße 11
79098 Freiburg
Germany
www.drwa.de

f/p design GmbH
Mauerkicherstraße 4
81679 München
Germany
www.f-p-design.com

Fuenfwerken Design AG
Taunusstraße 52
65183 Wiesbaden
Germany
www.fuenfwerken.com

FRACKENPOHL POULHEIM GMBH
Luxemburger Str. 72
50674 Köln
Germany
www.frackenpohl-poulheim.de

GDC Design
Kaiserstraße 168-170
90763 Fürth
Germany
www.gdc-design.de

H H Schultes Design Studio
Rodelbahnstraße 1
Germany
82256 Fürstenfeldbruck

i / i / d Institut für Integriertes Design
an der HfK Bremen
Am Speicher XI, Abtlg.7, Boden 3
28217 Bremen
Germany
www.iidbremen.de

identis GmbH
Bötzinger Straße 36
79111 Freiburg
Germany
www.identis.de

INOID DesignGroup
Reutlingerstraße 114
70597 Stuttgart
Germany
www.inoid.de

Interbrand
Kirchenweg 5
8008 Zürich
Switzerland
www.interbrand.com

MEDIADESIGN HOCHSCHULE
für Design und Informatik
Claudius-Keller-Straße 7
81669 München
Germany
www.mediadesign.de

mormedi
Plaza Republica Argentina 3
28002 Madrid
Spain
www.momedi.com

Nova Design Co., Ltd.
Tower C, 8F, No. 96, Sec. 1, Xintai 5th Rd.
Xizh City Taipei Country 221
Taiwan
www.e-novadesign.com

OCO-Design O. K. Nüsse
An der Kleimannbrücke 79
48157 Münster
Germany
www.oco-design.de

Olaf Hoffmann Industrial Design
Metzstraße 14 b
81667 München
Germany
www.olaf-hoffmann-design.de

Philips International BV Philips Design
Building HWD, Emmasingel 24
5600 MD Eindhoven
The Netherlands
www.philips.nl

Polvan Design Ltd.
Cemil Topuzlu cad. 79/2 Caddebostan
34170 Istanbul
Turkey
www.polvandesign.com

PROMOTIONAL iDEAS
Werbeagentur GmbH
Hessenring 76
61348 Bad Homburg
Germany
www.promotionalideas.de

rahe+rahe design
Konsul-Smidt-Straße 8c
28217 Bremen
Germany
www.rahedesign.de

Rokitta Produkt & Markenästhetik
Kölner Straße 38 a
45481 Mülheim an der Ruhr
Germany
www.rokitta.de

SCHOLZ & VOLKMER GmbH
Schwalbacher Straße 72
65183 Wiesbaden
Germany
www.s-v.de

Spirit Design –
Innovation and Brand GmbH
Hasnerstr. 123
1160 Wien
Austria
www.spiritdesign.com

Strategy & Marketing Institute
Management Consultants GmbH
Lange-Hop-Straße 19
30559 Hannover
Germany
www.strategy-institute.com

Taipei Base Design Center
1F.,No.4,Ln.176,Sec.1,Da'an Rd.,Da'an Dist.
Taipei City 106
Taiwan
www.asia-bdc.com

TRICON Design AG
Bahnhofstraße 26
72138 Kirchentellinsfurt
Germany
www.tricon-design.de

Ueberholz GmbH
Warndtstraße 7
42285 Wuppertal
Germany
www.ueberholz.de

UNIPLAN GmbH & Co. KG
Schanzenstraße 39 a/b
51063 Köln
Germany
www.uniplan.com

VDID/DDV
Markgrafenstraße 15
10969 Berlin
Germany
www.vdid.de

Weinberg & Ruf
Produktgestaltung
Martinsstraße 5
70794 Filderstadt
Germany
www.weinberg-ruf.de

PRIVATPERSONEN /
INDIVIDUAL MEMBERS

Thomas Bade
Im Knick 9
31655 Stadthagen
Germany

Olaf Barski
Oeder Weg 52-54
60318 Frankfurt
Germany

Thomas Biswanger
Probierlweg 47
85049 Ingolstadt
Germany

Christoph Böninger
Auerberg 1
83730 Fischbachau
Germany

Gerd Bulthaup
Chamissostraße 1
81925 München
Germany

Gerdum Enders
Terrasse 19
34117 Kassel
Germany

Holger Fricke
Loogestieg 19
20249 Hamburg
Germany

Andreas Gantenhammer
Meerbuscher Straße 64-78
40670 Meerbusch
Germany

Michael Grüter
Hainholzstraße 17
31558 Hagenburg
Germany

Annette Häfelinger
Johannisplatz 15
81677 München
Germany

Josef Hasberg
Lokenbach 8 – 10
51491 Overath
Germany

Bibs Hosak-Robb
Mendelssohnstraße 31
81245 München
Germany

Roy Huang
No. 4, Ln. 176, Sec. 1
Da`an Rd., Da`an Dist.
106 Taipei
Taiwan

Jens Korte
Frühlingsgarten 12
22297 Hamburg
Germany

Sebastian Le Peetz
Osterstraße 43 a
30159 Hannover
Germany

Hildegund Lichtwark
Lommertzweg 20
41569 Rommerskirchen
Germany

Johannes Loer
Luisenstraße 10
59379 Selm-Bork
Germany

Seyed Mansour
Pour Mohseni Shaikib
No.281- Yasseman Alley-Manzariyeh-Namjo St.
41936 Rasht
Iran

Tim Oelker
Venusberg 30
20459 Hamburg
Germany

Christoph Rohrer
Roecklplatz 3
80469 München
Germany

Karina Rudolph
Friedrichstraße 11
78050 VS – Villingen
Germany

HD Schellnack
Mintropstraße 61
45329 Essen
Germany

Eberhard Schlegel
Am Kapellenweg 4
88525 Dürmentingen
Germany

Michael Schlenke
Adlerstraße 16
41564 Kaarst
Germany

Gunnar Spellmeyer
c/o Fachhochschule Hannover Expo Plaza 2
30539 Hannover
Germany

Andreas Thierry
Arthur-Kutscher-Platz 1 / VII
80802 München
Germany

Peter Thonet
Michael-Thonet-Straße 1
35066 Frankenberg
Germany

Martin Topel
Fuhlrottstraße 10, Gebäude I, Ebene 16, Raum 76
42119 Wuppertal
Germany

Roland Wagner
Tullstraße 19
77933 Lahr
Germany

Gabriel Weber
Orffstraße 35
80637 München
Germany

 Nachwuchsförderung

iF concept design award 2014

Der iF concept design award zählt zu den größten Nachwuchswettbewerben weltweit. Bereits seit 2008 suchen wir jedes Jahr die 100 intelligentesten, innovativsten Konzepte aus allen Designdisziplinen. Studierende, Absolventen und Nachwuchsdesigner, die nicht länger als zwei Jahre im Beruf sind, können ihre Arbeiten einreichen. Insgesamt steht ein Preisgeld von EUR 30.000 zur Verfügung.

The iF concept design award is one of the most internationally renowned as well as biggest young designers competitions. Already since 2008 we are looking for the 100 most intelligent, most innovative concepts from all fields of design every year. Students and graduates that left university within the last two years are allowed to submit their entries. Altogether there is a prize money purse of EUR 30,000.

Hansgrohe Preis 2014

Bereits zum vierten Mal wird im Rahmen des iF concept design award der "Hansgrohe Preis" vergeben. Gefragt sind Konzepte rund um das Thema "Efficient Water Design". Das Preisgeld für diese Sonderauszeichnung beträgt EUR 5.000.

Already for the fourth time within the iF concept design award the "Hansgrohe Preis" is being held. This time we are looking for concepts touching the topic of "Efficient Water Design". The prize purse for this special prize is EUR 5,000.

iF concept design award 2014
GOLD SPONSORS

Support young talents

Haier Special Prize 2014

Erstmals wird innerhalb des iF concept design award der "Haier Special Prize 2014" ausgelobt. Hier können innovative Ideen rund um das Wettbewerbsthema "Wonderful Home Life - free your imagination on the elements of food, water and air" angemeldet werden. Das Preisgeld beträgt EUR 5.000.

For the first time it is possible to apply for the "Haier Special Prize 2014" within the iF concept design award. Here, innovative ideas related to the theme "Wonderful Home Life - free your imagination on the elements of food, water and air" can be submitted. The prize purse has the total amount of EUR 5,000.

IBDC 2014
International Bicycle Design Competition

In diesem internationalen Fahrrad-Wettbewerb, der seit 1997 jährlich stattfindet, werden innovative, visionäre Konzepte rund um das Thema „Bike" ausgezeichnet. EUR 21.500 Preisgeld wird an die herausragendsten Arbeiten verteilt.

This international bicycle competition, which has been held each year since 1997, rewards innovative, visionary concepts focusing on the topic of "bikes". EUR 21,500 in prize moneys will be distributed among the most outstanding entries.

BRONZE SPONSOR

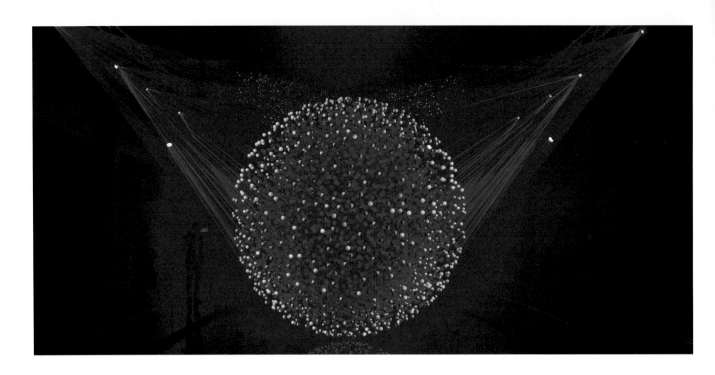

INDEX

MANUFACTURERS/CLIENTS

INDEX Manufacturer/Client

Symbole

1&1 Internet AG
Karlsruhe, Germany
page 46

3D Connexion
Munich, Germany
page 46

3M Taiwan Ltd.
Taipei, Taiwan
www.3M.com
page 46

4Qtrade GmbH
München, Germany
page 46

5 CUPS and some sugar
Berlin, Germany
www.5cups.de
page 46

© Südwestrundfunk
Baden-Baden, Germany
page 289

[supermusic]
Ratingen, Germany
supermusic.de
page 290

A

ABUS Security-Center
GmbH & Co. KG
Affing, Germany
www.abus.com
page 46

Acer Inc.
New Taipei City, Taiwan
www.acer.com
page 47

ACME Europe
Kaunas, Lithuania
www.acme.eu
page 47

Acrylic couture
Remagen-Kripp, Germany
www.acrylic-couture.com
page 48

Adam Opel AG
Rüsselsheim, Germany
www.opel.de
page 48

adidas AG
Herzogenaurach, Germany
www.adidas.de
page 48

Adlens Ltd.
Oxford, United Kingdom
www.adlens.com
page 48

admembers advertising GmbH
Duisburg, Germany
www.admembers.de
page 48

Aedilis
Vilnius, Lithuania
page 48

Aekyung Industry
Seoul, South Korea
page 49

A & E OPTICAL Ltd.
Shenzhen, China
page 49

Aerocrine AB
Solna, Sweden
www.aerocrine.com
page 49

Aesculap Akademie GmbH
Tuttlingen, Germany
page 49

AgênciaClick Isobar
São Paulo, Brazil
www.agenciaclick.com.br
page 49

Agentur am Flughafen
Altenrhein, Switzerland
page 49, 50

AhnLab
Seongnam-si, Gyeonggi-do,
South Korea
page 50

AIPHONE CO., Ltd.
Nagoya, Japan
www.aiphone.com
page 50

AiQ Smart Clothing Inc.
Taipei City, Taiwan
www.aiqsmartclothing.com
page 50

AIRACE ENTERPRISE Co., Ltd.
Tali Dist., Taichung City,
Taiwan
www.airace.com.tw
page 50, 51

Alape GmbH
Goslar, Germany
www.alape.com
page 51

Albers Inc.
New Taipei City, Taiwan
www.albers-creation.com
page 51

Albert Leuchten
Fröndenberg, Germany
www.gebr-albert.de
page 51

ALDI Supermercados
Barcelona, Spain
page 51

Alexander Tutsek-Stiftung
München, Germany
page 52

Alfred Galke GmbH
Gittelde, Germany
www.galke.com
page 52

Alfred Kärcher GmbH & Co. KG
Winnenden, Germany
page 52

Alfred Schladerer
Staufen im Breisgau,
Germany
page 52

AL-KO Geräte GmbH
Kötz, Germany
www.al-ko.de
page 52, 53

Allianz Deutschland AG
Unterföhring, Germany
page 53

ALNO AG
Pfullendorf, Germany
www.alno.de
page 53, 54

ALPEN-MAYKESTAG GmbH
Puch, Austria
www.alpenmaykestag.com
page 54

Alpnach
Norm-Schrankelemente AG
Alpnach-Dorf, Switzerland
www.alpnachnorm.ch
page 54

Alte Oper Frankfurt
Frankfurt a. M., Germany
www.alteoper.de
page 54

AMAZONEN-Werke
H. Dreyer GmbH & Co. KG
Hasbergen, Germany
www.amazone.de
page 55

Amnesty International
Frankfurt a. M., Germany
page 55

AM.PM
Berlin, Germany
www.ampm-world.com
page 55

Anders GmbH
Bad Staffelstein, Germany
page 55

Angelcare Monitors Inc.
Brossard, Québec, Canada
www.angelcare-monitor.com
page 56

Antec, Inc.
Fremont, CA,
United States of America
page 56

Appenzeller Schaukäserei
Stein AR, Switzerland
www.schaukaeserei.ch
page 277

Appenzeller Volkskunde-
Museum
Stein AR, Switzerland
www.appenzeller-museum-
stein.ch
page 277

Apple
Cupertino, CA,
United States of America
page 56, 57

Arbonia AG
Arbon, Switzerland
www.arbonia.ch
page 57

Arcadyan Technology
Corporation
Hsinchu, Taiwan
www.arcadyan.com
page 57

ARCA-SWISS
Besançon, France
page 57

Arçelik A. S.
Istanbul, Turkey
page 58, 59

Arco Contemporary
Furniture
Winterswijk, Netherlands
www.arco.nl
page 59

INDEX Manufacturer/Client

Ardagh Glass Limmared AB
Limmared, Sweden
www.ardaghgroup.com
page 59

Ardagh Group
Dublin, Ireland
www.ardaghgroup.com
page 59

Arda (Zhe Jiang)
Electric Co., Ltd.
Zhe Jiang Yong Kang, China
www.ardaappliance.com
page 59

Argent Alu
Kruishoutem, Belgium
www.argentalu.com
page 60

Arkoslight
Ribarroja del Turia, Spain
www.arkoslight.com
page 60

Armor Manufacturing Corp.
Taipei, Taiwan
www.cole-products.com
page 60

Armstrong DLW GmbH
Bietigheim-Bissingen,
Germany
www.armstrong.eu
page 60

Art Directors Club für
Deutschland (ADC) e.V.
Berlin, Germany
page 61

Artemide S. p. A.
Pregnana Milanese, Italy
www.artemide.com
page 61

Artweger GmbH & Co. KG
Bad Ischl, Austria
page 62

ASISTA Teile fürs Rad
GmbH & Co. KG
Leutkirch, Germany
www.asista.de
page 62

ASRock Incorporation
Tapei, Taiwan
www.asrock.com
page 62

Asstel ProKunde
Köln, Germany
page 62

ASUSTek Computer Inc.
Taipei, Taiwan
page 62 – 64

Ateliê Editorial
São Paulo, Brazil
page 64

ATELIER ALLURE
by THOMAS HAUSER
Wien, Austria
page 64, 65

Ateljé Lyktan AB
Åhus, Sweden
www.atelje-lyktan.se
page 65

AUDI AG
Ingolstadt, Germany
www.audi.de
page 65, 66

AUGUST
San Francisco, CA,
United States of America
www.august.com
page 66

AU Optronics Corporation
Hsinchu, Taiwan
www.auo.com
page 67

AUTHENTICS GmbH
Gütersloh, Germany
www.authentics.de
page 67

Autostadt GmbH
Wolfsburg, Germany
page 67

Autostadt GmbH
Wolfsburg, Germany
www.autostadt.de
page 67

Autostadt GmbH
Wolfsburg, Germany
www.autostadt.de/de/start
page 67

Avantgarde Acoustic GmbH
Lautertal-Reichenbach,
Germany
www.avantgarde-acoustic.de
page 68

AVerMedia
TECHNOLOGIES, Inc.
New Taipei City, Taiwan
page 68

AVEXIR Technologies
Corporation
Zhubei City, Taiwan
www.avexir.com
page 68

AWC AG
Köln, Germany
page 68

B

Backhaus Dries GmbH
Rüdesheim am Rhein,
Germany
page 68

Baliarne obchodu,
a.s. Proprad
Proprad, Slovakia
page 69

BALMUDA Inc.
Tokyo, Japan
page 69

bamboo
São Paulo, Brazil
bamboonet.com.br
page 69

Bang & Olufsen
Struer, Denmark
www.bang-olufsen.com
page 70

basalte
Merelbeke, Belgium
www.basalte.be
page 70

Bauerfeind AG
Zeulenroda-Triebes,
Germany
www.bauerfeind.com
page 70

Bayerische Motoren Werke AG
München, Germany
www.bmwgroup.com
page 70

Bayerisches
Staatsministerium
München, Germany
page 71

Beckhoff Automation GmbH
Verl, Germany
www.beckhoff.de
page 71

Bekina Boots N. V.
Kluisbergen, Belgium
page 71

Bel Epok GmbH
Köln, Germany
page 71

Bella Italia Weine
Stuttgart, Germany
www.bella-italia-weine.de
page 71

BEMOTEC GmbH
Reutlingen, Germany
www.bemotec.com
page 71

Bene AG
Waidhofen / Ybbs, Austria
www.bene.com
page 72

BenQ Corp.
Taipei, Taiwan
page 72

Berker GmbH & Co. KG
Schalksmühle, Germany
page 72

Bertelsmann SE & Co. KGaA
Gütersloh, Germany
www.bertelsmann.de
page 72

BET Aachen
Aachen, Germany
www.bet-aachen.de
page 72

betec Licht AG
Dachau, Germany
www.betec.de
page 73

BETO ENG. & MKTG. Co., Ltd.
Taichung, Taiwan
www.aplus-beto.com.tw
page 73

Betty Bossi AG
Zürich, Switzerland
www.bettybossi.ch
page 73

B & F Manufacture
GmbH & Co.KG
Schwäbisch Gmünd,
Germany
www.qlocktwo.com
page 73

bfs·batterie füllungs
systeme GmbH
Bergkirchen, Germany
page 73

Biccateca
Erechim, Brazil
www.biccateca.com.br
page 73

INDEX Manufacturer/Client

Bilfinger SE
Mannheim, Germany
www.bilfinger.de
page 74

Bischöfliche Aktion
Essen, Germany
www.adveniat.de
page 74

BlackBerry
Industrial Design Team
Waterloo, Canada
www.blackberry.com
page 74

Black Diamond
Equipment Limited
Salt Lake City, UT,
United States of America
www.
blackdiamondequipment.com
page 74

Blackmagic Design
Melbourne, Australia
www.blackmagic-design.com
page 75

BLANCO GmbH & Co. KG
Oberderdingen, Germany
www.blanco-germany.com/de
page 75

Bleijh Industrial Design
Amsterdam, Netherlands
www.sandwichbikes.com
page 75

blomus GmbH
Sundern, Germany
www.blomus.com
page 76

Bluebird
Seoul, South Korea
page 76

Bluewell Corporation
Gyeonggi-do, South Korea
page 76

Blume 2000
Norderstedt, Germany
page 76, 77

BMA Ergonomics
Zwolle, Netherlands
page 77

BMW AG
München, Germany
www.bmwgroup.com
page 77, 78

BMW Group
München, Germany
www.bmwgroup.com
page 78 – 80

BODUM AG
Triengen, Switzerland
page 80

Bolin Webb Ltd.
Uppingham,
United Kingdom
www.bolinwebb.com
page 81

BOMAG GmbH
Boppard, Germany
www.bomag.com
page 81

Bombardier Transportation
Hennigsdorf, Germany
page 81

Bosch Security Systems, Inc.
Burnsville, MI,
United States of America
us.boschsecurity.com
page 81

Bosch
Sicherheitssysteme GmbH
Grasbrunn, Germany
page 81, 82

Bosch Thermotechnik GmbH
Wetzlar, Germany
page 82

Bosch und Siemens
Hausgeräte GmbH
München, Germany
www.bsh-group.com
www.gaggenau.com
page 82

Braun Design
Kronberg i. T., Germany
www.braun.com
page 82

BRAUNWAGNER GmbH
Aachen, Germany
www.braunwagner.de
page 82

BRAVAT Plumbing Industrial
Co., Ltd.
GuangZhou, China
page 83

BREE Collection GmbH &
Co. KG
Isernhagen, Germany
www.bree.com
page 83

Bremer Feinkost GmbH &
Co. KG
Bremen, Germany
www.lappandfao.com
page 83

Brillengalerie
Bolay & Bolay OHG
Ulm, Germany
www.brillengalerie-bolay.de
page 83

BRITA GmbH
Taunusstein, Germany
www.brita.de
page 83

Britax Römer
Kindersicherheit GmbH
Ulm, Germany
www.britax-roemer.de
page 84

Brother Industries, Ltd.
Nagoya, Japan
page 84, 85

Brunner GmbH
Rheinau, Germany
www.brunner-group.com
page 85

BS and Co., Ltd.
Seoul, South Korea
page 85

BSH Home Appliances
(China) Co., Ltd.
Nanjing, China
page 85, 86

Bsize Inc.
Odawara, Kanagawa, Japan
www.bsize.com
page 87

Buddha Spirit club
Shanghai, China
page 87

BUFFALO Inc.
Nagoya, Japan
page 87

Bundesministerium
für Bildung
Berlin, Germany
www.bmbf.de
page 87

BURDA NEWS GROUP
Hamburg, Germany
page 91

Burkhardtsmaier
Untergruppenbach,
Germany
page 87

Bürositzmöbelfabrik
Dauphin
Offenhausen, Germany
page 88

Busch-Jaeger Elektro GmbH
Lüdenscheid, Germany
www.busch-jaeger.de
page 88

BUWON ELECTRONICS Co., Ltd.
Daegu, South Korea
page 88

B&W Group Ltd.
Steyning, United Kingdom
www.bwgroup.com
page 88

C

Calibre Style Ltd.
Taipei, Taiwan
page 88

Calor
Dublin 12, Ireland
www.calorgas.ie
page 88

Campus Verlag GmbH
Frankfurt am Main, Germany
www.campus.de
page 89

Canon Hongkong Co. Ltd.
Hong Kong
page 89

Canon Inc.
Tokyo, Japan
page 89, 90

CARREFOUR
Les Ulis, France
www.carrefour.com
page 90, 91

casamia
Seongnam, Gyeonggi-do,
South Korea
page 91

Cebien Co., Ltd.
Gyeonggi-do, South Korea
www.cebien.com
page 91

INDEX Manufacturer/Client

Celesio AG
Stuttgart, Germany
www.celesio.com
page 91

CE Lighting Ltd.
Shenzhen, China
www.landlite.com
page 91

CELLULAR GmbH
Hamburg, Germany
www.cellular.de
page 91, 92

Center for Medical Informatics
Seongnam-si, Gyeonggi-do, South Korea
www.snubh.org
page 92

Cerevo
Tokyo, Japan
www.cerevo.com
page 92

Changing Places Group
Cambridge, MA, United States of America
page 92

Cheil Germany GmbH
Schwalbach, Germany
www.cheil.com
page 92

ChenFangXiao Design International
Xiamen, China
www.cdi.hk
page 93

CHEN YU-MING
Kaohsiung, Taiwan
page 93

Chinatown Business Association
Singapore
page 93

ChoisTechnology Co., Ltd.
Incheon, South Korea
www.x-pointer.com
page 93

Chung-hea Myung-ga
Yongin-si, South Korea
page 93

Chungho Nais
Bucheon, Gyeonggi-Do, South Korea
www.eng.chungho.co.kr
page 93

Chung Hua University
Hsinchu, Taiwan
www.shkinetic.com
page 94

Cimax Design Engineering (Hong Kong) Ltd.
Shenzhen, China
www.libodesign.com
page 94

Cisco
Lysaker, Norway
www.cisco.com
page 94

CITIZEN Watch
Tokyo, Japan
www.citizen.co.jp
page 94

CJ CheilJedang
Seoul, South Korea
page 94, 95

C. Josef Lamy GmbH
Heidelberg, Germany
www.lamy.de
page 95

CLAAS KGaA mbH
Harsewinkel, Germany
www.claas.com
page 95, 96

Clever Pack
Rio de Janeiro, Brazil
www.cleverpack.com.br
page 96

Clormann Design GmbH
Penzing, Germany
www.clormanndesign.de
page 96

COMMAX
Kyunggi-Do, South Korea
www.commax.com
page 96

Commerz Real AG
Wiesbaden, Germany
www.commerzreal.com
page 96

Compal Electronic, Inc.
Taipei City, Taiwan
www.compal.com
page 97, 98

Compass pools Europe
Senec, Slovakia
page 98

CONARY ENTERPRISE CO., Ltd.
Taipei, Taiwan
page 99

Concord 2004, S. A.
Stadtsteinach, Germany
www.concord.es
page 99

Continental Reifen Deutschland GmbH
Hannover, Germany
page 99

Convotherm Elektrogeräte GmbH
Eglfing, Germany
www.convotherm.com
page 99

Coop Genossenschaft
Basel, Switzerland
www.coop.ch
page 99

Coordination Ausstellungs GmbH
Berlin, Germany
www.coordination-berlin.de
page 99

Cordivari S. R. L.
Morro D'Oro – TE, Italy
www.cordivaridesign.it
page 100

Coreana Cosmetics Co., Ltd.
Gyeonggi-do, South Korea
www.coreana.co.kr
page 100

Corning MobileAccess
Vienna, VA, United States of America
page 100

COUGAR
Tainan, Taiwan
www.cougar-world.com
page 100

COWAY
Seoul, South Korea
page 100, 101

COWON SYSTEMS Inc.
Seoul, South Korea
www.cowon.com
page 101

Croatian Designers Society
Zagreb, Republic of Croatia
www.dizajn.hr
page 101

Crown Gabelstapler GmbH & Co. KG
München, Germany
www.crown.com
page 101

cyber-Wear Heidelberg GmbH
Dossenheim, Germany
www.cyber-wear.de
page 128

D

DAEHAN A&C
Seoul, South Korea
www.daehananc.com
page 102

Daelim Dobidos
Incheon, South Korea
www.dlt.co.kr
page 102

Daimler AG
Stuttgart, Germany
www.daimler.com
page 102, 103

Dallmer GmbH & Co. KG
Arnsberg, Germany
www.dallmer.de
page 103

Daniela Fortinho Zilinsky
Blumenau, Brazil
page 103

DAPU tech. Co., Ltd.
Hangzhou, China
page 103

DART
Stuttgart, Germany
www.dartwork.de
page 103

DB Vertrieb GmbH
Frankfurt a. M., Germany
page 104

DDB Tribal Wien
Wien, Austria
page 104

DecoSlide Company A / S
Silkeborg, Denmark
www.decoslide.dk
page 104

Deff Corporation
Osaka City, Japan
www.deff.co.jp
page 104

Delica AG
Birsfelden, Switzerland
www.delizio.ch
page 105

INDEX Manufacturer/Client

De`Longhi Braun
Household GmbH
Neu-Isenburg, Germany
page 104

DELTA LIGHT N. V.
Wevelgem, Belgium
page 105

DENSO CORPORATION
Aichi, Japan
www.globaldenso.com/en
page 106

Department of Cultural
Affairs, Taipei City
Government
Taipei City, Taiwan
page 106

Design Board
Sofia, Bulgaria
www.designboard.com
page 106

Deutsche Lufthansa AG
Frankfurt, Germany
page 106

Deutsche Post DHL
Bonn, Germany
page 106

Deutscher Volkshochschul-
Verband e.V.
Bonn, Germany
page 106

Deutsches Historisches
Museum
Berlin, Germany
page 107

Deutsche Telekom AG
Bonn, Germany
www.telekom.de
page 107, 108

DEXINA AG
Böblingen, Germany
page 108

DIE LÖSUNG –
Strategie & Grafik
Memmingen, Germany
www.veraendertdiewelt.de
page 108

DIEPHAUS Betonwerk GmbH
Vechta, Germany
www.diephaus.de
page 108

digitalDigm
Seoul, South Korea
www.d2.co.kr
page 109

DIJIYA ENERGY SAVING
TECH. Inc.
Taoyuan County, Taiwan
www.dijiya.com
page 109

D-Link Corporation
Taipei, Taiwan
page 109

dm-drogerie markt GmbH &
Co. KG
Karlsruhe, Germany
page 109, 110

DocCheck AG
Köln, Germany
page 110

D.O.E.S
Seoul, South Korea
page 110

Domus Aurea GmbH
Stuttgart, Germany
www.domus-aurea.de
page 110

Dongdao Design Co., Ltd.
Beijing, China
www.dongdao.net
page 110

DORMA
Ennepetal, Germany
www.dorma.com
page 111

Dräger Safety AG & Co.
KGaA
Lübeck, Germany
www.draeger.com
page 111

Dr. Ing. h.c. F. Porsche AG
Stuttgart, Germany
www.porsche.com
page 111, 112

Dr. Oetker Nahrungsmittel KG
Bielefeld, Germany
page 112

Dropcam
San Francisco, CA,
United States of America
page 112

DURAN Group GmbH
Mainz, Germany
www.duran-group.com
page 112

Duravit AG
Hornberg, Germany
www.duravit.com
page 112, 113

DÜRR Ecoclean GmbH
Filderstadt, Germany
www.durr-ecoclean.com
page 113

Duscholux AG
Thun, Switzerland
www.duscholux.ch
page 113

Dutch House of
Representatives
Den Haag, Netherlands
page 113

Dyson
London, United Kingdom
page 114

E

Eastern Global Corporation
New Taipei City, Taiwan
www.eg.com.tw
page 114

Eco-Products Inc.
Boulder, CO,
United States of America
page 114

ECO Schulte GmbH & Co. KG
Menden, Germany
www.eco-schulte.de
page 114

Eczacibasi Yapi Gerecleri
Istanbul, Turkey
page 114, 115

Edifier International Ltd.
Admiralty, Hong Kong
www.edifier-international.com
page 115

Edimax Technology Co., Ltd.
New Taipei City, Taiwan
www.edimax.com
page 115

EEW
Energy from Waste GmbH
Helmstedt, Germany
www.eew-energyfromwaste.com
page 115

Ehinger Kraftrad
Katrin Oeding
Hamburg, Germany
page 115

E. I. Corporation
Hanam, South Korea
www.motherscornworld.com
page 115, 116

ELECOM Co., Ltd.
Chuoku, Osaka, Japan
www.elecom.co.jp
page 116

Electrolux AB
Stockholm, Sweden
page 116

Elektrobiker LTD & Co. KG
Wien, Austria
www.abikecalledquest.com
page 117

Elitegroup Computer
Systems Co., Ltd.
Taipei City, Taiwan
page 117

ELR Co., Limited
Selangor, Malaysia
www.elr-group.com
page 117

Elsner Elektronik GmbH
Gechingen, Germany
www.elsner-elektronik.de
page 117

ELV Elektronik AG
Leer, Germany
www.elv.de
page 117

Elvin Textile Company
Bursa, Turkey
www.elvinfabrics.com
page 117

E-Make Tools Co., Ltd.
Taichung, Taiwan
page 118

emart
Seoul, South Korea
page 118

Emart Company Ltd.
Seoul, South Korea
page 118

Emeco
Hanover, NH,
United States of America
www.emeco.net
page 118

English Egg
Seoul, South Korea
page 118

EOS Saunatechnik GmbH
Driedorf, Germany
www.eos-sauna.com
page 119

INDEX Manufacturer/Client

ERCO GmbH
Lüdenscheid, Germany
www.erco.com
page 119

erfi – Ernst Fischer
GmbH & Co. KG
Freudenstadt, Germany
www.erfi.de
page 119

ERMAKSAN
Bursa, Turkey
www.ermaksan.com.tr
page 119

Ernst A. Geese GmbH
Henstedt-Ulzburg, Germany
page 232

ETAP N. V.
Malle, Belgium
www.etaplighting.com
page 119

Eton Corporation
Palo Alto, CA,
United States of America
page 120

Evangelische
Hochschule Berlin
Berlin, Germany
page 120

Eva Solo A / S
Måløv, Denmark
www.evasolo.com
page 120

Exklusiv Hauben
Gutmann GmbH
Mühlacker, Germany
page 120

Exped AG
Zürich, Switzerland
www.exped.com
page 121

F

Fabbian Illuminazione S. p. A.
Resana, Italy
www.fabbian.com
page 121

facts and fiction GmbH
Köln, Germany
www.factsfiction.com
page 121

Fancom
Panningen, Netherlands
www.fancom.nl
page 121

FARM
Rio de Janeiro, Brazil
www.farmrio.com.br
page 122

Fashion group Hyung ji
Seoul, South Korea
www.hyungji.co.kr
page 121

Favour Light Company
Limited
San Po Kong, Hong Kong
www.favourlight.com
page 122

FAXSON Co., Ltd.
Taichung City, Taiwan
page 122

FERMAX ELECTRONICA S. A. U.
Valencia, Spain
www.fermax.com
page 122

Ferromatik Milacron
Malterdingen, Germany
www.ferromatik.com
page 122

Festo AG & Co. KG
Esslingen, Germany
www.festo.com
page 123, 124

FESTOOL Group
GmbH & Co. KG
Wendlingen, Germany
page 124

Fiat Automóveis S/A
Betim, Brazil
www.fiat.com.br
page 49

Fiat Group Automobiles
Germany AG
Frankfurt a. M., Germany
page 124

Fiat Group Automobiles
Germany AG
Frankfurt a. M., Germany
www.fiat.de
page 124

Fibar Group
Poznan, Poland
www.fibaro.com
page 124

fiftyfifty
Düsseldorf, Germany
page 124

Fisher & Paykel
Auckland, New Zealand
www.fisherpaykel.co.nz
page 125

Fissler GmbH
Idar-Oberstein, Germany
www.Fissler.de
page 125

Fitbit Inc.
San Francisco, CA,
United States of America
www.fitbit.com
page 125

Fleurance Nature
Fleurance, France
page 125

flexi-Bogdahn International
GmbH & Co. KG
Bargteheide, Germany
www.flexi.de
page 126

FLEYE
Hedehusene, Denmark
www.fleye.dk
page 126

FLÖTOTTO
Systemmöbel GmbH
Rietberg, Germany
www.floetotto.de
page 126

FLOWAIR Sp.j.
Gdynia, Poland
www.flowair.com
page 126

FLUVIA CONCEPT, S. L. U.
Madrid, Spain
www.fluvia.com
page 126

FLUX-GERÄTE GmbH
Maulbronn, Germany
www.flux-pumpen.de
page 127

FontanaArte S. p. A.
Corsico, Italy
www.fontanaarte.com
page 127

Food Industry Research and
Development Institute
Hsinchu, Taiwan
www.firdi.org.tw
page 127

Ford Werke GmbH
Köln, Germany
www.ford.com
page 128

Foshan Nanhai Chevan
Kayl, Luxembourg
www.vanguardworld.com
page 128

Fraport AG
Frankfurt a. M., Germany
www.fraport.de
page 128

frog
San Francisco, CA,
United States of America
page 128

Frost A / S
Hadsten, Denmark
www.frostdesign.dk
page 128

FUJIFILM Corporation
Minato-ku, Tokyo, Japan
www.fujifilm.com
page 128, 129

Fuji Heavy Industries Ltd.
Tokyo, Japan
page 129

FUJITSU Ltd.
Kawasaki, Japan
www.fujitsu.com/global
page 130

function Technology AS
Bergen, Norway
page 130

FUN FACTORY GmbH
Bremen, Germany
www.funfactory.de
page 130

G

Gaggenau Hausgeräte GmbH
München, Germany
www.gaggenau.com
www.gaggenau.com/de
page 130

Game Technologies S. A.
Poznań, Poland
www.game-technologies.com
page 131

Garden City Publishers
Taipei, Taiwan
www.gardenct.pixnet.net
page 131

GD Midea Air-Conditioning
Equipment Co., Ltd.
FoShan, China
page 131

INDEX Manufacturer/Client

Geberit International AG
Jona, Switzerland
page 131, 132

GemVax & Kael
Bundang-gu, Seongnam-si,
Gyeonggi-do, South Korea
page 132

Germanwings GmbH
Köln, Germany
www.germanwings.com
page 132

GfG
Dortmund, Germany
www.gasmessung.de
page 132

Giesecke & Devrient GmbH
München, Germany
www.gi-de.com
page 132

GIGANTEX
Changhua County, Taiwan
page 133

Gigaset
Communications GmbH
München, Germany
www.gigaset.com
page 133

Giotto's Ind. Inc.
New Taipei City, Taiwan
www.giottos.com
page 133

GIXIA GROUP Co.
Taipei, Taiwan
page 133, 134

GMC-I Messtechnik GmbH
Nürnberg, Germany
www.gossenmetrawatt.com
page 134

GN Netcom A/S
Ballerup, Denmark
www.jabra.com
page 134

Goang Yuh Shing Ltd.
Taipei, Taiwan
page 134

GOGANG Aluminium Co., Ltd.
Seoul, South Korea
page 134

GÖKSEL ARAS – OFİSLINE
Sivas, Turkey
www.ofisline.com.tr
page 135

Golden Sun
Home Products Ltd.
Hong Kong, China
www.goldensun.com.hk
page 135

Goodbaby
Child Products Co., Ltd.
Utrecht, Netherlands
page 135

Google Deutschland GmbH
Hamburg, Germany
page 135

Google, Inc.
Mountain View, CA,
United States of America
page 135

GP Acoustics (UK) Ltd.
Maidstone Kent,
United Kingdom
www.kef.com
page 136

Grey Düsseldorf GmbH
Düsseldorf, Germany
www.grey.de
page 136

Griffwerk GmbH
Blaustein, Germany
www.griffwerk.de
page 136

Grohe AG
Düsseldorf, Germany
www.grohe.com
page 136

Guangdong Ganlanz
Microwave Oven and
Electrical Appliances
Manufacturing Co., Ltd.
Shunde, China
page 137

Guangdong Midea Kitchen
Guang Dong, China
page 137

Guangdong OPPO Mobile
Telecommunications Corp., Ltd.
Dongguan, China
www.oppo.com
page 137

Guangzhou Shirui
Electronics Co., Ltd.
Guangzhou, China
www.iseewo.com
page 138

Guangzhou Times
Property Group
Guangzhou, China
page 138

Guilin Yangshuo Nice View
Guilin, Guangxi, China
page 138

Gunitech Corp.
Qionglin Township, Hsinchu
County, Taiwan
www.gunitech.com
page 138

GWA Bathrooms & Kitchens
Sydney, Australia
www.gwagroup.com.au
page 139

GWI Gemeindewerke
Ismaning
Ismaning, Germany
www.hallenbad-ismaning.de
page 139

H

Haier Group
Qingdao, China
page 139, 140

Hair O'right International Corp.
Taoyuan County, Taiwan
page 140

Hallingers
Schokoladenmanufaktur
Landsberg a. Lech, Germany
www.hallingers.de
page 140

Hälssen & Lyon
Hamburg, Germany
page 141

HAMMER CASTER Co., Ltd.
Higashi-Osaka City, Osaka,
Japan
www.hammer-caster.co.jp
page 141

Hangulhwa
Seoul, South Korea
page 141

Hangzhou ROBAM
Appliances Co., Ltd.
Hangzhou, China
www.robam.com
page 141

Hangzhou Teague
Technology Co., Ltd.
HangZhou, China
www.teague.net.cn
page 141

Hankook Tire Co., Ltd.
Seoul, South Korea
www.hankooktire.com
page 142

Hansa Metallwerke AG
Stuttgart, Germany
www.hansa.de
page 142

Hansgrohe SE
Schiltach, Germany
www.hansgrohe.com
www.axor-design.com
page 142, 143

HAPPY IN COMPANY
Seoul, South Korea
www.sysmax.co.kr
page 143

HARIO Co., Ltd.
Tokyo, Japan
www.hario.jp
page 143

Harman International
Industries, Inc.
Northridge, CA,
United States of America
www.harman.com
page 143 – 145

Hartware
MedienKunstVerein
Dortmund, Germany
page 146

Hasegawa Kogyo
Osaka, Japan
www.hasegawa-kogyo.co.jp
page 146

Haufe-Lexware
GmbH & Co. KG
Freiburg, Germany
www.lexware.de
page 146

Hausmann's Flughafen
Gastronomie
Hamburg, Germany
www.hausmanns-frankfurt.de
page 146

Haworth GmbH
Ahlen, Germany
page 146

Axor Starck Organic

Follow your **Head** and your *Heart*

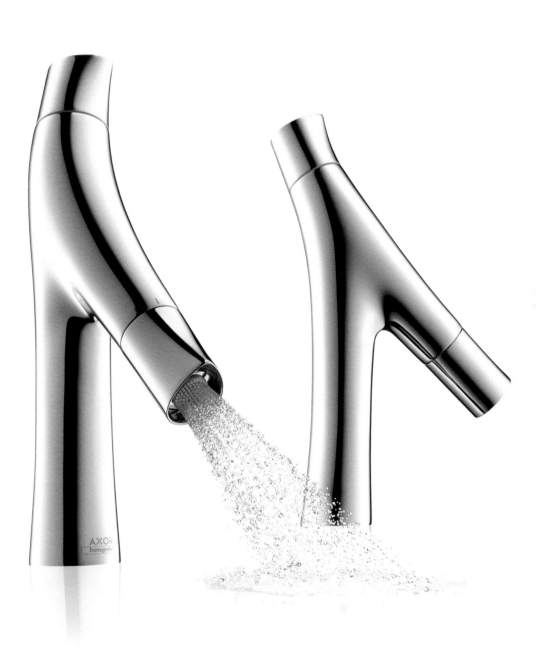

▸ Wie viel Herz und wie viel Verstand stecken in Axor Starck Organic? Außergewöhnlich viel: ein so noch nie dagewesener funktionaler Armaturenbrausestrahl, ein sparsamer Wasserverbrauch von 3,5 l/min, ein vollkommen neues Bedienkonzept. Mehr zur neuen Badkollektion, die Sie das Wasser auf ganz neue Art erleben lässt: **head-and-heart.com**

H.-D. SCHUNK GmbH & Co.
Mengen, Germany
www.schunk.com
page 147

Heine/Lenz/Zizka
Projekte GmbH
Frankfurt a. M., Germany
www.hlz.de
page 147

Helge Nyberg AB
Ulricehamn, Sweden
page 147

Hello AG
München, Germany
www.hello-muenchen.de
page 147

HengLu Myoung Gift Co., Ltd.
Hangzhou, China
page 103

Hengyang 1more Electronic
Technology Co., Ltd.
Shenzhen, China
page 147

Herman Miller
Wiltshire, United Kingdom
www.hermanmiller.com
page 148

HERM. STIESING KG BREMEN
Bremen, Germany
www.stiesing.de
page 147

Hettich
Marketing- und Vertriebs
Kirchlengern, Germany
www.hettich.com
page 148

HEWI Heinrich Wilke GmbH
Bad Arolsen, Germany
www.hewi.de
page 148

Hewlett-Packard
San Diego, CA,
United States of America
page 149

Hewlett-Packard
San Diego, CA,
United States of America
www.hp.com
page 149, 150

Hewlett-Packard
Sunnyvale, CA,
United States of America
www.hp.com
page 149, 150

Hewlett-Packard Company
Netanya, Israel
www.hp.com
page 150

Hewlett-Packard Company
Palo Alto, CA,
United States of America
page 135, 150

Hilti Corporation
Schaan, Liechtenstein
page 151, 152

Hisense International Co., Ltd.
Qingdao, China
page 152

Hitachi Appliances, Inc.
Tokyo, Japan
page 152

Hitachi Koki Co., Ltd.
Hitachinaka-City, Ibaraki
Pref., Japan
www.hitachi-koki.co.jp
page 152, 153

HOBART GmbH
Offenburg, Germany
page 153

Hoffmann GmbH
Qualitätswerkzeuge
München, Germany
www.hoffmann-group.com
page 154

Honeywell Safety Products
Lübeck, Germany
www.honeywellsafety.com
page 154

Honeywell Safety Products
Smithfield, RI,
United States of America
www.honeywellsafety.com
page 154

Hong Kong Resort
Company Limited
Discovery Bay, Hong Kong
www.aubergediscoverybay.com
page 154

HORIBA, Ltd.
Kyoto, Japan
www.horiba.com
page 154

HSK Duschkabinenbau KG
Olsberg, Germany
page 155

HTC
New Taipei City, Taiwan
page 155

Huawei Device Co., Ltd.
Shenzhen, China
www.consumer.huawei.com
www.huaweidevice.com
page 155, 156

Huawei Technologies Co., Ltd.
Shanghai, China
page 156

Hubei Insun Cinema
Film Co., Ltd.
Wuhan, China
page 156

Hultafors Group AB
Hultafors, Sweden
www.hultaforsgroup.com
page 156

HUMAN Gesellschaft für
Biochemica
Wiesbaden, Germany
page 157

HUNTER DOUGLAS
EUROPE B. V.
Rotterdam, Netherlands
www.hunterdouglasgroup.com
page 157

HÜPPE GmbH
Bad Zwischenahn, Germany
www.hueppe.com
page 157

Hurom L. S. Co., Ltd.
Gimhae-si, South Korea
www.hurom.com
page 157

hw.design GmbH
München, Germany
page 158

Hyundai Amco
Seoul, South Korea
www.hyundai-amco.co.kr
page 158

Hyundai Card
Seoul, South Korea
page 158

Hyundai Motor Company
Frankfurt a. M., Germany
www.hyundai.com, www.
innocean.com
page 159

Hyundai Motor Company
Seoul, South Korea
page 158

Hyundai Motor Group
Seoul, South Korea
page 159

I

icontrols
Sung-Nam-Si, Gyung-Gi-Do,
South Korea
www.icontrols.co.kr
page 159

IDEA DO IT
Daegu, South Korea
www.ideadoit.com
page 159

IDEAL STANDARD
INTERNATIONAL BVBA
Zaventem, Belgium
www.idealstandard.com
page 159

ID+IM Design Lab., KAIST
Daejeon, South Korea
www.idim.kaist.ac.kr
page 160

ID Infinity Limited
Hong Kong
page 160

iDtools B. V.
Delft, Netherlands
www.idtools.eu
page 160

ifm electronic
Essen, Germany
www.ifm.com
page 160

igus® GmbH
Köln, Germany
www.igus.de
page 160, 161

iHealth Lab Inc.
Mountain View, CA,
United States of America
www.ihealthlabs.com
page 161, 162

Iittala Group Oy Ab
Helsinki, Finland
www.iittala.com
page 162, 163

ILSHIN AUTOCLAVE
Daejeon, South Korea
page 163

ilusyd
Yongin, South Korea
page 163

Impossible
Wien, Austria
www.the-impossible-
project.com
page 163

INDEX Manufacturer/Client

INDGROUP
Daegu, South Korea
page 164

Industrial Technology
Research Institute
Hsinchu, Taiwan
www.itri.org.tw
page 164, 168, 169

Inge Ingwersirup
München, Germany
www.die-inge.de
page 164

Ingo Maurer GmbH
München, Germany
www.ingo-maurer.com
page 164

INNODevice, Inc.
Gyeung gi do, South Korea
www.innodevice.com
page 164

Inspur (Beijing) Electronic
Beijing, China
www.inspur.com
page 164

Institute for Information
Industry
Taipei City, Taiwan
page 165

Insuline Medical Ltd.
Petach Tikva, Israel
www.insuline-medical.com
page 165

Interbrand Korea
Seoul, South Korea
page 165

intergram
Seoul, South Korea
www.intergram.co.kr
page 165

Interprint GmbH
Arnsberg, Germany
www.interprint.de
page 165

Interstil
Steinhagen, Germany
www.interstil.de
page 165

Interstuhl Büromöbel
GmbH & Co. KG
Meßstetten-Tieringen,
Germany
www.interstuhl.de
page 166

Intra lighting
Šempeter pri Gorici, Slovenia
www.intra-lighting.com
page 166

IP44 Schmalhorst
GmbH & Co. KG
Rheda-Wiedenbrück,
Germany
www.IP44.de
page 166

IPSA CO., Ltd.
Tokyo, Japan
www.ipsa.co.jp
page 166

IRIVER
Seoul, South Korea
www.iriver.com
page 167

iRobot Corporation
Bedford, PY,
United States of America
www.irobot.com
page 167

ISSEY MIYAKE Inc.
Tokyo, Japan
page 167, 168

IT DESIGN
Seoul, South Korea
page 168

item Industrietechnik GmbH
Solingen, Germany
www.item24.com
page 168

J

JADO
Zaventem, Belgium
www.jado.de
page 169

Jaquar & Company Pvt. Ltd.
Gurgaon, Haryana, India
www.artize.in
page 169

Jardin De Jade
Shanghai, China
page 169

Jawbone Industrial Co., Ltd.
New Taipei City, Taiwan
www.jawbone.com.tw
page 170

JBX SYSTEM
Taichung, Taiwan
page 170

JEILTECH
Incheon, South Korea
page 170

Jeju Special Self-governing
Province Development Corp.
Jeju, South Korea
www.jpdc.co.kr
page 170

Jenoptik Robot GmbH
Monheim am Rhein,
Germany
www.jenoptik.com
page 170

Jewelry Shalom
Chiyoda-ku Tokyo, Japan
www.shalom-marunouchi.com
page 171

JIA Inc.
Hong Kong
page 171

J. Morita Mfg. Corp.
Kyoto, Japan
page 171

JOBY Inc.
San Francisco, CA,
United States of America
www.joby.com
page 171

JOMOO Kitchen & Bath
Appliances Co., Ltd.
Xiamen, China
www.jomoo.com.cn
page 171, 172

Joseph Joseph
London, United Kingdom
www.josephjoseph.com
page 172

JoyaTech
Xiamen, China
page 172

Joy Industrial Co., Ltd.
Taichung Hsien, Taiwan
page 173

JUDO
Wasseraufbereitung GmbH
Winnenden, Germany
www.judo.eu
page 173

Just Mobile Ltd.
Taichung, Taiwan
www.just-mobile.com
page 173

K

KABA
Villingen-Schwenningen,
Germany
page 173

Kähler Design A / S
Næstved, Denmark
www.kahlerdesign.com
page 174

Kamstrup A / S
Skanderborg, Denmark
www.kamstrup.com
page 174

Kan & Lau Design
Consultants
Kowloon Tong, Hong Kong
page 174

Kanton Solothurn
Solothurn, Switzerland
page 174

Kaohsiung
Museum of Fine Art
Kaohsiung, Taiwan
page 174

Karl Knauer KG
Biberach / Baden, Germany
www.karlknauer.de
page 175

KBS
Seoul, South Korea
www.kbs.co.kr
page 175

KD One Co., Ltd.
Seoul, South Korea
page 175

Kenwood Limited
Hampshire, United Kingdom
page 175

Kesseböhmer GmbH
Bad Essen, Germany
www.kesseboehmer.de
page 175

KEUCO GmbH & Co. KG
Hemer, Germany
www.keuco.de
page 175, 176

KfW Stiftung
Frankfurt a. M., Germany
www.kfw-stiftung.de
page 176

Kia Motors Corporation
Frankfurt a. M., Germany
www.kiamotors.com
page 176

KIA Motors Corporation
Seoul, South Korea
www.kiamotors.com
page 158

KIDU
Daejeon, South Korea
www.kidu.co.kr
page 176

KILIAN Tableting GmbH
Köln, Germany
page 176

KIMS HOLDINGS
Busan, South Korea
page 176

Kinnasand GmbH
Westerstede, Germany
www.kinnasand.com
page 177

KION GROUP AG
Wiesbaden, Germany
www.kiongroup.com
page 177

KircherBurkhardt GmbH
Berlin, Germany
page 177

Kleemann GmbH
Göppingen, Germany
www.kleemann.info
page 177

Klever Mobility Inc.
New Taipei City, Taiwan
www.klever-mobility.com
page 177

KMS TEAM GmbH
München, Germany
www.kms-team.de
page 177

KMW
Hwaseong-si, Gyeonggi-do,
South Korea
page 178

Knabenkantorei Basel
Münchenstein, Switzerland
page 178

Koga B. V.
Heerenveen, Netherlands
www.koga.com
page 178

Kohler
China Investment Co., Ltd.
Shanghai, China
www.kohler.com.cn
page 178

Kohler Mira Ltd.
Cheltenham, United
Kingdom
page 178

Koike Rice Store
Tokyo, Japan
page 179

Koleksiyon Mobilya
San. A. Ş.
Istanbul, Turkey
www.koleksiyon.com.tr
page 179

Kölner Freiwilligen Agentur
Köln, Germany
www.koeln-freiwillig.de
page 179

KOLON
Gumi, South Korea
page 179

KONE Corporation
Hyvinkää, Finland
www.kone.com
page 179

Koninklijke Gazelle N. V.
Dieren, Netherlands
page 180

Konrad Hornschuch AG
Weißbach, Germany
www.skai.com
www.hornschuch.com
page 180

Konstantin Slawinski
Siegen, Germany
www.konstantinslawinski.com
page 180

Konzerthaus Berlin
Berlin, Germany
www.konzerthaus.de
page 180

Kose Corporation
Tokyo, Japan
www.kose.co.jp
page 180

Kösel GmbH & Co. KG
Altusried-Krugzell, Germany
page 232

Krones AG
Neutraubling, Germany
page 181

kt media hub
Seoul, South Korea
page 181

Kuhn Rikon AG
Rikon, Switzerland
www.kuhnrikon.ch
page 181

KUKA AG
Augsburg, Germany
page 181

Kumho Tire Co.
Yongin-si, Gyeonggi-do,
South Korea
page 182

Kunstdünger GmbH
Silandro, Italy
www.kdmarket.it
page 295

Kuoni Reisen Holding AG
Zürich, Switzerland
page 182

Kuoyenting
Taichung City, Taiwan
page 182

Kverneland Group
Kerteminde, Denmark
www.kvernelandgroup.com
page 182

KWH Mirka Ltd.
Jeppo, Finland
www.mirka.com
page 183

KYE SYSTEMS Corp. (Genius)
Taipei Hsien, Taiwan
www.geniusnet.com
page 183

KYOCERA Corporation
Kyoto, Japan
global.kyocera.com
page 183

kyowon L & C
Seoul, South Korea
page 183

L

LAB.C
Seoul, South Korea
page 183

LACO
Uhrenmanufaktur GmbH
Pforzheim, Germany
www.laco.de
page 184

Landesanglerverband
Brandenburg e.V.
Nuthetal OT Saarmund,
Germany
www.landesanglerverband-bdg.de
page 184

Landor and Associates
Hamburg, Germany
www.landor.com
page 184

Langenscheidt GmbH & Co. KG
München, Germany
www.langenscheidt.de
page 184

Leadtek Research Inc.
Taipei, Taiwan
page 184

LED Linear GmbH
Neukirchen-Vluyn, Germany
www.led-linear.de
page 184

LEICHT Küchen AG
Waldstetten, Germany
www.leicht.de
page 185

Leifheit AG
Nassau / Lahn, Germany
www.soehnle.de
page 185

LEMKEN GmbH & Co. KG
Alpen, Germany
page 185

Lenovo
Morrisville, NC,
United States of America
page 185, 186

Lenovo (Beijing) Ltd.
Beijing, China
page 186

Lernwelt Wolfpassing
Wolfpassing, Austria
page 187

Lesmo GmbH & Co. KG
Düsseldorf, Germany
page 187

leyess
Seoul, South Korea
www.lyanature.com
page 187

LFI Photographie GmbH
Hamburg, Germany
page 187

IM NOTFALL KEINE KOMPROMISSE
ECHTE PROFIS BRAUCHEN PROFI-AUSRÜSTUNG!

WELTWEIT EXZELLENTER RUF
Ausgereifte Technologie, einfache Bedienführung, große Mobilität und absolute Zuverlässigkeit unter Extrembedingungen: unverwechselbare Kennzeichen der PRIMEDIC™-Defibrillatoren. Kein Zufall, dass die überaus robusten Geräte in der Reanimations- und Notfallmedizin weltweit einen exzellenten Ruf genießen.

QUALITÄT ALS PRINZIP
Professionelle Rettungskräfte setzen auf Qualität, denn sie müssen sich auf ihre Ausrüstung verlassen können. Konstruiert für den Einsatz durch Notärzte und Rettungsdienste, verbindet der Defibrillator maximale Ausstattung mit größtmöglicher Zuverlässigkeit. Damit im Einsatz keine wertvolle Zeit verloren geht.

www.primedic.com

PRIMEDIC™
Saves Life. Everywhere.

INDEX Manufacturer/Client

LG Electronics Inc.
Seoul, South Korea
www.lg.com
page 187 – 192

LG Hausys
Seoul, South Korea
www.lghausys.com
page 192

LG innotek
Ansan, South Korea
www.lginnotek.com
page 192

Libratone A/S
Herlev, Denmark
www.libratone.com
page 192

Licon mt GmbH & Co. KG
Laupheim, Germany
www.licon.com
page 192

Liebherr
Kirchdorf / Iller, Germany
www.liebherr.com
page 192

Life Safety Products BV
Zoetermeer, Netherlands
www.lifehammerproducts.com
page 193

Lifetrons Switzerland AG
Dicken, Switzerland
www.lifetrons.ch
page 193

LIGHTperMETER
Waregem, Belgium
www.lightpermeter.be
page 193

Lightweight Containers
Den Helder, Netherlands
www.keykeg.com
page 193

Linde AG
München, Germany
www.linde.de
page 194

Linde
Material Handling GmbH
Aschaffenburg, Germany
www.linde-mh.de
page 194

LINE
Seongnam-si, Gyeonggi-do,
South Korea
www.line.me
page 194

LI-NING (China) Sports
Goods Co.,
Beijing, China
page 194

Lino Manfrotto + Co. S. p. A.
Bassano del Grappa, Italy
page 194

Lin's Ceramics Studio
New Taipei City, Taiwan
www.taurlia.com
page 195

linshaobin design
Shantou, China
www.linshaobin.com
page 195

LIULIGONGFANG
Shanghai, China
www.liuliliving.com
page 195

LM Audit und Tax GmbH
München, Germany
www.lmat.de
page 195

LMS Instruments B. V.
Breda, Netherlands
www.lmsintl.com
page 195

LOBTEX Co., Ltd.
Osaka, Japan
www.lobtex.co.jp
page 195

Loewe AG
Kronach, Germany
www.loewe.tv
page 196

LOGICDATA GmbH
Deutschlandsberg, Austria
www.logicdata.at
page 196

Logitech
Newark, CA,
United States of America
www.logitech.com
page 196

Logitech Europe
Lausanne, Switzerland
www.logitech.com
page 197

Logitech Gateway
Newark, NJ,
United States of America
page 197

Lord Benex
International Co., Ltd.
Taichung City, Taiwan
www.lordbenex.com
page 197

LOTTE Confectionery
Seoul, South Korea
page 198

Louis Poulsen Lighting A / S
København K, Denmark
page 198

LPKF AG
Garbsen, Germany
www.lpkf.com
page 198

LSA International
Sunbury on Thames, United Kingdom
page 198

lumini
São Paulo, Brazil
www.lumini.com.br
page 198, 199

LUNAR
München, Germany
page 199

LUNAR
San Francisco, CA,
United States of America
page 199

Luxuni GmbH
Leer, Germany
page 199

LYA NATURE
Seoul, South Korea
page 200

Lytro, Inc.
Mountain View, CA,
United States of America
page 200

M

Madeblunt Ltd.
Auckland, New Zealand
www.bluntumbrellas.com
page 200

Made in China by Boldº
Rio de Janeiro, Brazil
page 200

Magic Bubble
Beijing, China
page 200

Makita Werkzeug GmbH
Ratingen, Germany
page 201

Maltani lighting
Seoul, South Korea
www.taewon.co.kr
page 201

Mammut Sports Group AG
Seon, Switzerland
www.mammut.ch
page 201

Marcopolo S/A
Caxias do Sul, Brazil
www.marcopolo.com.br
page 201

Marina Brodersby
Brodersby, Germany
page 201

MARTINELLI LUCE S. p. A.
Lucca, Italy
www.martinelliluce.it
page 202

MASTER S. p. A.
Martellago, Italy
www.mastermotion.eu
page 202

Matrix Audio Ltd.
ONT, Canada
www.matrixaudio.com
page 202

MAVIC S. A. S.
Annecy Cedex 9, France
page 202

MAVIN TECHNOLOGY Inc.
Chupei City, Taiwan
www.mavintec.com
page 202

MAXU Technology Inc.
Jiangsu, China
page 202

McDonald's
Wien, Austria
page 203

McDonald's Deutschland Inc.
München, Germany
page 203

Medion AG
Essen, Germany
page 203

INDEX Manufacturer/Client

Medtronic
Minneapolis, MN,
United States of America
www.medtronic.com
page 204

Melitta SystemService
GmbH & Co. KG
Minden, Germany
www.melitta.de
page 204

Menlo Systems GmbH
Martinsried, Germany
www.menlosystems.com
page 204

MENNEKES Elektrotechnik
GmbH & Co. KG
Kirchhundem, Germany
www.mennekes.de
page 204

Mercury Marine
Fond Du Lac, WI,
United States of America
www.mercurymarine.com
page 204

Merit Co., Ltd.
Daejeon, South Korea
page 204

Merry Electronics Co., Ltd.
Taichung, Taiwan
page 205

Merten GmbH
Wiehl, Germany
www.merten.de
page 205

Metrax GmbH
Rottweil, Germany
www.primedic.com
page 205

METRO AG
Düsseldorf, Germany
page 205

Metso Corporation
Helsinki, Finland
page 205

Metso Lindemann
Düsseldorf, Germany
www.metso.com
page 206

Metso Minerals, Inc.
Tampere, Finland
page 206

Meyer & Meyer Holding
GmbH & Co. KG
Osnabrück, Germany
page 206

Michel Mercier
Tel Aviv, Israel
www.michelmercier.com
page 206

Microsoft
Redmond, WA,
United States of America
page 206

Miele & Cie. KG
Gütersloh, Germany
www.miele.de
page 207

MiGlaz Design Co., Ltd.
Nonthaburi, Thailand
www.miglaz.com
page 207

Milla & Partner
Stuttgart, Germany
www.milla.de
page 207

MINOX GmbH
Wetzlar, Germany
www.minox.com
page 207

miscea GmbH
Berlin, Germany
www.miscea.com
page 207

Misfit Wearables
Corporation
Redwood City, CA,
United States of America
www.misfitwearables.com
page 207

MiTAC International Corp.
Taipei, Taiwan
www.mitac.com
page 208

Mitsubishi Electric
Corporation
Tokyo, Japan
page 208

Mitsubishi Electric Europe B. V.
Ratingen, Germany
www.mitsubishielectric.de
page 209

MITSUBISHI PENCIL Co., Ltd.
Tokyo, Japan
page 209

MOBICA+
Eckental, Germany
www.mobicaplus.de
page 209

MOBIC
Daegu, South Korea
page 209

Moldex-Metric AG & Co. KG
Walddorfhäslach, Germany
www.moldex-europe.com
page 209

Mosarte
Tijucas – SC, Brazil
www.mosarte.com.br
page 209

Motorola Solutions
Schaumburg, IL,
United States of America
www.motorolasolutions.com
page 210

MTT Integration Corp.
George Hill, Anguilla
page 210

muehlhausmoers
Berlin, Germany
www.muehlhausmoers.com
page 210

MuKK GmbH
Münster, Germany
page 210

Museum für
Gegenwartskunst Siegen
Siegen, Germany
www.mgk-siegen.de
page 211

Museum Villa Stuck
München, Germany
www.villastuck.de
page 211

M. Water Co., Ltd.
Bangkok, Thailand
www.sprinkle-th.com
page 211

N

Nanjing HuaQi
Visual Design Co., Ltd.
Nanjing, China
www.huaqiart.com
page 211

NAOS Design
Karlsruhe, Germany
www.naos.de
page 211

National Museum of
Contemporary Art, Korea
Gwacheon, South Korea
www.mmca.go.kr
page 211

National Museum of
Denmark
Copenhagen, Denmark
page 212

NAVER Corp.
Seongnam-si, Gyeonggi-do,
South Korea
page 212

NEC Display Solutions Ltd.
Kanagawa, Japan
page 212

Nedap Livestock
Management
Groenlo, Netherlands
en.nedap-
livestockmanagement.com
page 212

Neolab convergence
Seoul, South Korea
page 213

Nescafé Dolce Gusto
Vevey, Switzerland
page 213

new&able
Karlsruhe, Germany
page 213

New Zealand Opera
Auckland, New Zealand
www.nzopera.com
page 213

NEXENTIRE Corporation
Yangsan Kyungnam,
South Korea
www.nexentire.com
page 213

NHN Entertainment
Seongnam-si, South Korea
page 213

Nico Bats, Fabian Bremer
Frankfurt / Berlin / Leipzig,
Germany
page 214

Nike, Inc.
Beaverton, OR,
United States of America
page 214

NIKE SPORTS KOREA CO., Ltd.
Seoul, South Korea
page 215

Nikon Corporation
Tokyo, Japan
page 215

Nilfisk-Advance A / S
Hadsund, Denmark
page 215

Nils Holger Moormann GmbH
Aschau im Chiemgau,
Germany
www.moormann.de
page 215

Nilvia Lab
Mussolente, Italy
www.nilvia.it
page 215

Ningbo Ledeshi Electric
Equipment C
Ningbo, China
www.ledeshi.com
page 216

Nippon Columbia Co., Ltd.,
Tokyo, Japan
www.columbia.jp
page 216

Nissan Global
Yokohama, Kanagawa,
Japan
page 216

NISSHO JITSUGYO CO., Ltd.
Osaka, Japan
page 216

No 2 Records
Frankfurt a. M., Germany
page 216

Nokia
Beijing, China
page 217

Nokia
Espoo, Finland
page 217

Nokia
Espoo, Finland
www.nokia.com
page 216

Nokia
London, United Kingdom
page 218

Nokia
San Diego, CA,
United States of America
www.nokia.com
page 217

Nokia Corporation
Beijing, China
page 218

Nokia Corporation
Espoo, Finland
page 218

NOMOS Glashütte / S. A.
Glashütte, Germany
www.nomos-glashuette.com
page 218

Norbert Woll GmbH
Saarbrücken, Germany
page 219

Norbert Woll GmbH
Saarbrücken, Germany
www.woll-cookware.com
page 219

NOVAREVO
Incheon, South Korea
www.novarevo.net
page 219

Novescor GmbH
Mainz, Germany
www.melicena.com
page 219

NTT DOCOMO, Inc.
Tokyo, Japan
page 219

NTT DOCOMO, Inc.
Tokyo, Japan
www.nttdocomo.co.jp
page 219

NWMD GmbH
Berlin, Germany
www.nwmd.de
page 219

Nya Nordiska Textiles GmbH
Dannenberg, Germany
page 220

O

OAXIS Holdings Pte. Ltd.
Singapore
page 220

Occhio GmbH
München, Germany
www.occhio.de
page 221

Odenwälder
Kunststoffwerke
Gehäusesysteme GmbH
Buchen, Germany
www.okw.com
page 221

Ofa Bamberg GmbH
Bamberg, Germany
www.ofa.de
page 221

Okamura Corporation
Yokohama, Japan
www.okamura.co.jp
page 221

Oki Seki Co., Ltd.
Kanagawa, Japan
www.okiseki.com/index.html
page 221

OLIGO Lichttechnik GmbH
Hennef, Germany
www.oligo.de
page 222

OLYMP GmbH & Co. KG
Stuttgart, Germany
www.olymp.de
page 222

Olympus Corporation
Tokyo, Japan
www.olympus.com
page 222

Olympus Deutschland GmbH
Hamburg, Germany
www.olympus.de
page 222

OMRON HEALTHCARE Co., Ltd.
Kyoto, Japan
www.healthcare.omron.co.jp
page 222, 223

One Plus Partnership Ltd.
Hong Kong
www.onepluspartnership.com
page 223

Optrel AG
Wattwil, Switzerland
www.optrel.com
page 223

Ora ïto Mobility
Paris, France
page 224

ordinarypeople
Seoul, South Korea
www.ordinarypeople.kr
page 224

Osram GmbH
München, Germany
www.osram.com
page 224

Ótima
São Paulo, Brazil
www.otima.com
page 224

Otto Bock
Mobility Solutions GmbH
Königsee, Germany
www.ottobock.de
page 225

Otto Ganter GmbH & Co. KG
Furtwangen, Germany
www.ganter-griff.de
page 225

Oventrop GmbH & Co. KG
Olsberg, Germany
www.oventrop.de
page 225

Oxylane
Lille, France
www.decathlon.com
page 225

Oxylane
Villeneuve d'Ascq, France
www.artengo.com
page 225

OZAKI INTERNATIONAL
Co., Ltd.
New Taipei City, Taiwan
www.ozakiverse.com
page 226

Oz Estratégia + Design
São Paulo, Brazil
www.ozdesign.com.br
page 226

P

Pacific Market International
Amsterdam, Netherlands
page 226

Palamides GmbH
Renningen, Germany
www.palamides.de
page 226

Panasonic Corporation
Osaka, Japan
www.panasonic.net
page 227 – 230

Panasonic Deutschland
Hamburg, Germany
page 230

PANTECH CO., Ltd.
Seoul, South Korea
page 230

Paper Pleasure
Penzing, Germany
www.paper-pleasure.de
page 230

FORM FOLLOWS FUNCTION

Louis Sullivan, amerikanischer Architekt, 1856-1924

pinox Der Thermostat

Die Liebe zum Detail:
Sie ist wichtig, um sich wohlzufühlen.

Inspiriert durch die Formensprache moderner Einrichtungen setzt der „pinox" Thermostat besondere Akzente. Funktional und formvollendet macht er an Heizkörpern eine gute Figur. Der Griff ermöglicht eine leichte und präzise Temperatureinstellung mit „Fingerspitzengefühl". Der „pinox" wurde mit zahlreichen Design Preisen ausgezeichnet.

Oventrop bietet vorteilhafte Lösungen für Heizen, Kühlen und Trinkwasser.

OVENTROP GmbH & Co. KG
Paul-Oventrop-Straße 1, D-59939 Olsberg
www.oventrop.de

INDEX Manufacturer/Client

Parador GmbH & Co. KG
Coesfeld, Germany
www.parador.de
page 230

Partec GmbH
Görlitz, Germany
www.partec.com
page 230, 231

PAX
Computer Technology Co., Ltd.
Shenzhen, China
page 231

Pegatron Corporation
Beitou District, Taipei City,
Taiwan
page 231

Pegatron Corporation
Taipei, Taiwan
page 231

pester pac automation GmbH
Wolfertschwenden,
Germany
www.pester.com
page 231

Peyer Graphic GmbH
Leonberg, Germany
www.peyergraphic.de
page 232

PF Concept International B. V.
Roelofarendsveen,
Netherlands
www.pfconcept.com
page 232, 233

Pfeffersack & Soehne
Koblenz, Germany
www.pfeffersackundsoehne.de
page 233

PFLITSCH GmbH & Co. KG
Hückeswagen, Germany
www.pflitsch.de
page 233

Phiaton
Irvine, CA,
United States of America
www.phiaton.com
page 233

PHILIPPI GmbH
Hamburg, Germany
www.philippi.com
page 233

philosys Co., Ltd.
Gunsan-si, Jeollabuk-do,
South Korea
www.philosys.com
page 233

Phoenix Tapware
Bayswater, Australia
www.phoenixtapware.com.au
page 234

Piano Viabilità Polo
Luganese
Lugano, Switzerland
page 234

Pica-Marker
Kirchehrenbach, Germany
www.pica-marker.com
page 234

Plantronics
Santa Cruz, CA,
United States of America
page 234, 235

Plateau Wine Trading
Seoul, South Korea
page 235

Plus Eyewear Limited
Sanpokong, Hong Kong
www.pluseyewear.com
page 235

POLA Inc.
Tokyo, Japan
www.pola.co.jp
page 235

P.O.S. Co., Ltd.
Tokorozawa, Saitama, Japan
www.hygge-watches.com
page 235

PROBAT-Werke von Gimborn
Maschinenfabrik GmbH
Emmerich am Rhein,
Germany
www.probat.com
page 235

Promate Technologies
Shenzhen, China
www.promate.net
page 236

P & T – Paper & Tea
Berlin, Germany
page 236

Pulmuone
Health & Living Co., Ltd.
Seoul, South Korea
page 236

Q

Qinhetang
International Co., Ltd.
Xitun Dist., Taichung City,
Taiwan
page 236

Qisda Corporation
Taipei, Taiwan
page 236, 237

QNAP Systems, Inc.
New Taipei City, Taiwan
www.qnap.com
page 237

Quanta Computer Inc.
Tao Yuan, Taiwan
www.quantatw.com
page 237

QUECHUA
Domancy, France
www.quechua.com
page 237

QUMEI FURNITURE GROUP
Co., Ltd.
Beijing, China
www.qumei.com
page 237

R

Raiffeisenbank
St. Gallen, Switzerland
page 238

Random House
Auckland, New Zealand
www.randomhouse.co.nz
page 238

Raum B Architektur
Zürich, Switzerland
www.raumb.ch
page 238

Raytrix GmbH
Kiel, Germany
www.raytrix.de
page 238

Reckitt Benckiser
Slough, United Kingdom
www.rb.com
page 238

Red Indians Publishing
GmbH & Co. KG
Reutlingen, Germany
www.ramp-magazin.de
page 238, 239

Reflect Inc.
Seoul, South Korea
page 239

Reiko Kitora
Tokyo, Japan
page 239

Relations Inc.
Shibuya-ku, Tokyo, Japan
www.relationsgroup.co.jp
page 239

Renata Rubim
Porto Alegre – RS, Brazil
www.renatarubim.com.br
page 239

RenQing Technology Co., Ltd.
Shenzhen, China
www.ihave.hk
page 240

repaBAD GmbH
Wendlingen, Germany
www.repabad.com
page 240

REWE Markt GmbH
Köln, Germany
page 240

RICOH
Yokohama-shi, Japan
www.ricoh.co.jp
page 240, 241

Robert Bosch
Hausgeräte GmbH
München, Germany
www.bosch-hausgeraete.de
page 241 – 243

Robert Welch Designs Ltd.
Chipping Campden,
United Kingdom
page 244

rosconi GmbH
Weilburg, Germany
www.rosconi.com
page 244

Röthlisberger Kollektion
Gümligen, Switzerland
www.roethlisberger.ch
page 244

ROX Asia Consultancy Ltd.
Emsdetten, Germany
page 175

Royal Philips Electronics
Eindhoven, Netherlands
www.philips.com
page 244 – 252

RTI Sports GmbH
Koblenz, Germany
www.rtisports.de
page 252

ruwido austria GmbH
Neumarkt a. W., Austria
www.ruwido.com
page 252

INDEX Manufacturer/Client

Ryohin Keikaku Co., Ltd.
Tokyo, Japan
www.muji.co.jp
page 252, 253

S

Saint-Gobain Weber GmbH
Düsseldorf, Germany
www.sg-weber.de
page 253

Sal. Oppenheim jr. & Cie. KGaA
Köln, Germany
page 253

Salvatore Tiles & Tapetes
Lindolfo Collor, Brazil
www.salvatoreminuano.com.br
page 253

Samet
Istanbul, Turkey
www.samet.com.tr
page 253

Samkwang Glass Co., Ltd.
Seoul, South Korea
www.glasslock.co.kr
page 254

Samsung Cheil Industries
Gyeonggi-Do, South Korea
www.cii.samsung.com
page 254

Samsung Electronics Co., Ltd.
Gyeonggi-do, South Korea
page 255

Samsung Electronics Co., Ltd.
Seoul, South Korea
www.samsung.com
page 255 – 260

Samsung Everland
Yongin, South Korea
www.everland.com
page 261

Samsung SDS Co., Ltd.
Seoul, South Korea
page 261

Sankyo-Alumi Company
Toyama, Japan
www.alumi.st-grp.co.jp
page 261

SANTOX
Löffingen, Germany
www.santox.com
page 261

SANYO Electric Co., Ltd.
Osaka, Japan
page 261, 262

Sattler GmbH
Heiningen, Germany
www.sattler-lighting.com
page 262

Saturnbath. Co., Ltd.
Seoul, South Korea
www.saturn.co.kr
page 262

Saunalux GmbH
Products & Co. KG
Grebenhain, Germany
www.saunalux.de
page 262

Sayuri Studio, Inc.
Tokyo, Japan
www.ss-studio.com
page 262

Schaffner EMV AG
Luterbach, Switzerland
www.schaffner.com
page 262

Schauspielhaus Bochum
Bochum, Germany
www.schauspielhausbochum.de
page 263

Schindelhauer Bikes
Berlin, Germany
www.schindelhauerbikes.de
page 263

SCHMIDHUBER
München, Germany
www.schmidhuber.de
page 263

Schneider Electric
São Paulo, Brazil
www.schneider-electric.com
page 263

Scholz & Volkmer GmbH
Wiesbaden, Germany
www.s-v.de
page 263

Schönek GmbH & Co. KG
Nittenau, Germany
page 264

Schuberth GmbH
Magdeburg, Germany
www.schuberth.com
page 264

Schüco International KG
Bielefeld, Germany
www.schueco.com
page 264

SCIENCE INT'L CO., Ltd.
Taipei, Taiwan
www.mr-sci.com
page 264

seepex GmbH
Bottrop, Germany
page 265

Seiko Epson Corporation
Suwa-shi, Nagano-ken, Japan
www.epson.com
page 265

SELF Electronics Co., Ltd.
Ningbo, China
www.self-electronics.com
page 265

Sementes Ipiranga
Alto Taquari, Brazil
page 265

Sempio Foods Company
Seoul, South Korea
www.sempio.com
page 265

Sennheiser
Communications A/S
Ballerup, Denmark
www.senncom.com
page 266

Sennheiser
electronic GmbH & Co. KG
Wedemark, Germany
www.sennheiser.com
page 266

Sense
Beijing, China
page 266

sentiotec GmbH
Regau, Austria
www.sentiotec.com
page 266

SERCOMM
Taipei, Taiwan
www.sercomm.com
page 267

Sergio Bertti
Gramado, Brazil
page 267

Shanghai United Imaging
Healthcare Co., Ltd.
Shanghai, China
page 267

Shang Yih
Interior Design Co., Ltd.
Taipei, Taiwan
page 267

Shengtai Brassware Co., Ltd.
Chang Hua, Taiwan
www.justime.com
page 268

Shenji Group Kunming
Machine Tool
Kunming, China
page 268

Shenzhen Baojia Battery
Tech Co., Ltd.
Shenzhen, China
www.mipow.com
page 268, 269

Shenzhen Breo
Technology Co., Ltd.
Shenzhen, China
www.breocare.com
page 270

Shenzhen Puremate
Technology Co., Ltd.
Shenzhen, China
www.puremate.cn
page 270

Shenzhen QVOD
Technology Co., Ltd.
Shenzhen, China
www.qvod.com
page 270

Shenzhen Rapoo
Technology Co., Ltd.
Shenzhen, China
page 270

Shenzhen Shenghongxing
Technology Co., Ltd.
Shenzhen, China
page 270

Shenzhen Uoshon
Communication Technology
Co., Ltd.
Shenzhen, China
www.uoshon.com
page 270

Shimano Inc.
Osaka, Japan
page 271

Shimano Inc.
Sakai City, Japan
www.shimano.com
page 271

Ship and Ocean Industries
R & D Center
New Taipei, Taiwan
www.soic.org.tw
page 271

SiB Co., Ltd.
Tokyo, Japan
www.urbanutility.com
page 272

SICK Engineering GmbH
Ottendorf-Okrilla, Germany
www.sick.com
page 272

SieMatic Möbelwerke
Löhne, Germany
www.siematic.de
page 272

Siemens AG
Erlangen, Germany
www.siemens.com
page 272

Siemens AG
München, Germany
page 272

Siemens AG, Industry Sector
Nürnberg, Germany
page 272, 273

Siemens
Audiologische Technik GmbH
Erlangen, Germany
www.siemens.com/hoergeraete
page 273

Siemens
Electrogeräte GmbH
München, Germany
www.siemens-hausgeraete.de
page 273, 274

SIG Combibloc
Linnich, Germany
www.sig.biz
page 274

Sign Architecture & Interior
Design Co., Ltd.
Taipei, Taiwan
www.signarchi.com.tw
page 274

Silicon Power Computer &
Communications Inc.
Taipei, Taiwan
www.silicon-power.com
page 274

Similor AG
Laufen, Switzerland
www.laufen.com
page 274

simple GmbH
Köln, Germany
www.simple.de
page 275

Singapore Airlines
Singapore
www.singaporeair.com
page 275

Sirona Dental Systems GmbH
Bensheim, Germany
www.sirona.de
page 275

Skantherm Wagner
GmbH & Co. KG
Oelde, Germany
www.skantherm.de
page 275

SK
Seoul, South Korea
page 275

SKF Maintenance Products
Nieuwegein, Netherlands
page 275

SK Planet
Seoul, South Korea
www.skplanet.com
page 276

SKS metaplast Scheffer-
Klute GmbH
Sundern, Germany
www.sks-germany.com
page 276

SKYLOTEC GmbH
Neuwied, Germany
www.skylotec.de
page 276

Smaller International Co., Ltd.
Taipei, Taiwan
page 276

S.M. Entertainment
Seoul, South Korea
page 276, 277

Smith Optics
Ketchum, ID,
United States of America
www.smithoptics.com
page 277

SNA Europe [Sweden] AB
Enköping, Sweden
www.bahco.com
page 277

SO Appenzeller Käse GmbH
Appenzell, Switzerland
www.appenzeller.ch
page 277

SodaStream
Mt. Laurel, NJ,
United States of America
www.sodastreamusa.com
page 278

Sodastream Ltd.
Ben Gurion Airport, Israel
page 278

SOLARLUX Aluminium
Systeme GmbH
Bissendorf, Germany
www.solarlux.de
page 278

Sollosbrasil
Princesa, Brazil
page 278

Soma
San Francisco, CA,
United States of America
www.drinksoma.com
page 278

SOMA spol. s. r. o.
Lanskroun, Czech Republic
www.soma-eng.com
page 279

SonoScape Co., Ltd.
Shenzhen, China
page 279

Sonos Europe B. V.
Hilversum, Netherlands
page 279

SONY Corporation
Tokyo, Japan
www.sony.net/design
page 279 – 283

Sony EMCS Corporation
Tokyo, Japan
page 283

Sony Enterprise Co., Ltd.
Chuo-ku, Tokyo, Japan
page 283

Sony Mobile
Communications Inc.
Tokyo, Japan
page 283, 284

Sparkassenverband
Niedersachsen
Hannover, Germany
www.svn.de
page 284

Sputnik Engineering AG
Biel, Switzerland
page 284

S. Siedle & Söhne
Furtwangen, Germany
www.siedle.de
page 285

STABILO International GmbH
Nürnberg, Germany
page 285

Stadler Form
Aktiengesellschaft
Zug, Switzerland
page 285

Städtische Werke
Magdeburg GmbH
Magdeburg, Germany
www.sw-magdeburg.de
page 285

Stadt Schorndorf
Schorndorf, Germany
www.schorndorf.de
page 285

Stadt Waldkirch
Waldkirch, Germany
www.identis.de
page 286

STAEDTLER Mars GmbH &
Co. KG
Nürnberg, Germany
www.staedtler.de
page 286

STAHLWILLE
Wuppertal, Germany
www.stahlwille.de
page 286

State Government
of Minas Gerais
Belo Horizonte, Brazil
page 286

Stechert
Stahlrohrmöbel GmbH
Wilhermsdorf, Germany
www.stechert.de
page 286

Steelcase S. A.
Schiltigheim, France
www.steelcase.fr
page 287

talantone
CREATIVE GROUP

ADDRESS Suite #601, 512-8, Sinsa-Dong, Gangnam-Gu, Seoul 135-887, Korea
CONTACT INFORMATION +82-2-515-5761
serendeepity@talantone.com
www.talantone.com

1. **Cheil Industries 2012 Corporate Report** Winner (2014 iF Communication Design Award), Grand (2013 ARC Awards), Gold (2012 LACP Vision Awards) 2. **Korean Air 2011 Annual Report** Platinum (2011 LACP Vision Awards, Global Top 30) 3. **Cheil Industries 2010 Corporate Report** Grand (2011 ARC Awards), Gold (2010 LACP Vision Awards) 4. **Kia Motors 2012 Annual Report** Gold (2012 LACP Vision Awards), Bronze (2013 ARC Awards) 5. **Dong-A Pharmaceutical 2012 Annual Report** Gold (2012 LACP Vision Awards), Silver (2013 ARC Awards) 6. **Kumho Tires Product Catalogues** 7. **Kumho Tires 2013 Calendar** 8. **KDB Financial Group 2012 Annual Report** Gold (2012 LACP Vision Awards), Bronze (2013 ARC Awards) 9. **Cheil Industries Product Brochures**

SteelSeries ApS
Valby, Denmark
www.steelseries.com
page 287

STEIL KRANARBEITEN
GmbH & Co. KG
Trier, Germany
www.steil-kranarbeiten.de
page 287

Steinel GmbH
Herzebrock-Clarholz,
Germany
www.steinel.de
page 287

Stelton A / S
Copenhagen K, Denmark
www.stelton.com
page 287

Stiefelmayer-Lasertechnik
GmbH & Co
Denkendorf, Germany
page 288

Stiftung
Luthergedenkstätten
Lutherstadt Wittenberg,
Germany
page 288

STI Group
Greven, Germany
www.sti-group.com
page 288

STI Group
Lauterbach, Germany
www.sti-group.com
page 288

stilhaus AG
Rothrist, Switzerland
page 288

STILL GmbH
Hamburg, Germany
page 288

Storck Bicycle GmbH
Idstein, Germany
www.storck-bicycle.de
page 289

Stryker Leibinger
GmbH & Co. KG
Freiburg, Germany
www.stryker.com
page 289

Sudhaus GmbH & Co. KG
Iserlohn, Germany
www.sudhaus.de
page 289

Sunrise Medical
GmbH & Co. KG
Malsch, Germany
www.sunrisemedical.de
page 290

Sunstar Inc.
Osaka, Japan
jp.sunstar.com
page 290

Suunto Oy
Vantaa, Finland
www.suunto.com
page 290

Svakom Design USA, Limited
Newark, DE,
United States of America
page 290

Swareflex GmbH
Vomp, Austria
www.swareflex.com
page 291

Swegon AB
Kvänum, Sweden
www.swegon.com
page 291

sygonix GmbH
Nürnberg, Germany
www.sygonix.de
page 291

Symrise AG
Holzminden, Germany
www.symrise.com
page 291

SYZYGY Deutschland GmbH
Frankfurt a. M., Germany
www.syzygy.de
page 291

T

tado° GmbH
München, Germany
www.tado.com
page 291

TAEJU Industry. Inc.,
Seongdong-gu,Seongsu
2-ga 3-dong, South Korea
www.taeju.kr
page 292

Taiwan Design Center
Taipei, Taiwan
www.tdc.org.tw/en
page 292

Taiwan Noodle House
Beijing, China
page 292

Taiyo Toryo Co., Ltd.
Tokyo, Japan
www.maskingcolor.com
page 292

Takagi Co., Ltd.
Kitakyushu, Japan
page 292, 293

talsee
Hochdorf, Switzerland
www.talsee.ch
page 293

Tangible Space
Seoul, South Korea
page 293

Tatsuuma-Honke Brewing
Company, Ltd.
Hyogo, Japan
www.hakushika.co.jp
page 293

TCS AG
Triberg, Germany
www.carus-concepts.com
page 293, 294

TEAM 7 Natürlich Wohnen
GmbH
Ried im Innkreis, Austria
www.team7.at
page 294

teamandproducts GmbH
München, Germany
page 294

Teataster Group Holdings
Limited
Taipei, Taiwan
page 294

Teatulia Organic Teas
Denver, CO,
United States of America
www.teatulia.com
page 294

Technicolor
Issy Les Moulineaux, France
www.technicolor.com
page 294

Telefónica Germany
GmbH & Co. OHG
München, Germany
page 295

Tempaline AG
Riehen, Switzerland
www.tempaline.com
page 295

Tetra Pak
Modena, Italy
www.tetrapak.com
page 295

Teuco Guzzini S. p. A.
Montelupone, Italy
www.teuco.com
page 295

Teunen Konzepte GmbH
Geisenheim, Germany
page 295

Theben AG
Haigerloch, Germany
page 296

The Brand Union GmbH
Hamburg, Germany
www.thebrandunion.de
page 296

Thermaltake
Technology Co., Ltd.
Taipei, Taiwan
www.thermaltake.com.tw
page 296

THIMM Verpackung
GmbH & Co. KG
Northeim, Germany
www.thimm.de
page 296

Thinkware
Seongnam-si, South Korea
page 296

Thonet GmbH
Frankenberg, Germany
www.thonet.de
page 297

Thule AB
Malmö, Sweden
www.thule.com
page 297

Thule Group
Malmö, Sweden
www.thulegroup.com
page 297

Thüringer Energie AG
Erfurt, Germany
www.thueringerenergie.de
page 297

INDEX Manufacturer/Client

ThyssenKrupp Uhde GmbH
Dortmund, Germany
www.thyssenkrupp-uhde.de
page 297

Tianjin Art Museum
Tianjin, China
page 298

Timesco Ltd.
Basildon, United Kingdom
www.timesco.com
page 298

Tobias Grau GmbH
Rellingen, Germany
www.tobias-grau.com
page 298

Tombow Pencil Co., Ltd.
Tokyo, Japan
www.tombow.com
page 298

Tommee Tippee
Cramlington,
United Kingdom
www.tommeetippee.com
page 299

TomTom International B. V.
Amsterdam, Netherlands
www.tomtom.com
page 299

tonwelt
professional media GmbH
Berlin, Germany
www.tonwelt.com
page 299

TopGun / Tforce / Ubike
Yuan Lin Town, Changhua,
Taiwan
www.ubike-tech.com
page 299

Top Victory Electronics
(Taiwan) Co., Ltd.
New Taipei City, Taiwan
page 299

Tornos S. A.
Moutier, Switzerland
www.tornos.com
page 300

Toshiba
Research Triangle Park, NC,
United States of America
page 300

Toshiba Corporation
Tokyo, Japan
page 300

Toshiba Home Appliances
Corporation
Tokyo, Japan
page 300, 301

TOTO (China) Co., Ltd.
Chao Yang District, Beijing,
China
page 301, 302

TOTO Europe GmbH
Düsseldorf, Germany
page 302

TOTO Ltd.
Tokyo, Japan
www.toto.co.jp
page 301, 302

TOWA Co., Ltd.
Tokyo, Japan
www.towa-inc.co.jp
page 302

Town of Arlington
Financial Office
Arlington, VA,
United States of America
page 302

TP-LINK Technologies Co., Ltd.
Nanshan, Shenzhen, China
www.tp-link.com
page 302

TREE PLANET
Seoul, South Korea
www.treepla.net
page 302, 303

TRIAND Inc.
Sagamihara,Kanagawa,
Japan
www.triand.jp
page 240

TRILUX GmbH & Co. KG
Arnsberg, Germany
www.trilux.de
page 303

Trippen A. Spieth,
M. Oehler GmbH
Berlin, Germany
www.trippen.com
page 303

TRO GmbH
Düsseldorf, Germany
www.tro.de
page 303, 304

true fruits GmbH
Bonn, Germany
page 304

TSUBAKI KABELSCHLEPP GmbH
Wenden-Gerlingen,
Germany
www.kabelschlepp.de
page 304

TUNTUN English
Seoul, South Korea
page 304

Tupperware France S. A.
Joue-Les-Tours, France
www.tupperware.com
page 304

U

Uhlmann Pac-Systeme
GmbH & Co. KG
Laupheim, Germany
www.uhlmann.de
page 304

Ultimate Ears
Irvine, CA,
United States of America
page 305

Uncommon LLC
Chicago, IL,
United States of America
page 305

Unikia S. A.
Asker, Norway
page 305

Unitron
Kitchener, Canada
unitron.com
page 305

Uponor Group
Vantaa, Finland
www.uponor.com
page 305

urbn; interaction
Berlin, Germany
www.urbn.de
page 306

Ustar Biotechnologies
(Hangzhou) Ltd.
Hangzhou, China
www.weibo.com
page 306

V

Vacheron Constantin
Plan-les-Ouates, Switzerland
page 306

Vaillant GmbH
Remscheid, Germany
page 306, 307

Van den Weghe
Zulte, Belgium
www.lapris.be
page 307

vangard
Düsseldorf, Germany
page 307

Van Hoecke
Sint-Niklaas, Belgium
www.taor.com
page 307

VELUX A / S
Hørsholm, Denmark
www.velux.com
page 307

Venjakob
Maschinenbau GmbH
Rheda-Wiedenbrück,
Germany
www.venjakob.de
page 307

Verdom
Maceió, Brazil
page 308

Vermop Salmon GmbH
Gilching, Germany
page 308

Vestel Beyaz Eşya
San. ve Tic. A. Ş.
Manisa, Turkey
page 308

Viessmann Werke
GmbH & Co. KG
Allendorf / Eder, Germany
www.viessmann.com
page 308, 309

Vifa Denmark A/S
Viborg, Denmark
www.vifa.dk
page 309

VIKING GmbH
Langkampfen / Kufstein,
Austria
www.viking-garden.com
page 309

Villeroy & Boch AG
Mettlach, Germany
www.villeroy-boch.de
page 309

Ares Combo

Incredibly Simplified & Elegant By Design

Bow Tie | Full Touch | Dead Front | CFM | SMARTSPACE®

 global consumer design

Graphite*

INDEX Manufacturer/Client

VITEO GmbH
Graz, Austria
www.viteo.com
page 309

VitrA Karo
Istanbul, Turkey
www.vitra.com.tr
page 309, 310

Vogel's Products B. V.
Eindhoven, Netherlands
page 310

Völkl Sports GmbH & Co. KG
Straubing, Germany
www.voelkl.com
page 310

Volkswagen AG
Wolfsburg, Germany
www.volkswagen.de
www.volkswagenag.com
page 310, 311

Volvo Construction
Equipment
Changwon, South Korea
www.volvoce.com
page 312

Volvo Trucks
Gothenburg, Sweden
www.volvotrucks.com
page 312

VS Vereinigte
Spezialmöbelfabriken
Tauberbischofsheim,
Germany
page 312

V-ZUG AG
Zug, Switzerland
www.vzug.ch
page 312

W

Wacom Company Ltd.
Tokyo, Japan
www.wacom.com
page 313

Wagner GmbH
Langenneufnach, Germany
page 313

Wagner System GmbH
Lahr, Germany
page 313

Walter Knoll AG & Co. KG
Herrenberg, Germany
www.walterknoll.de
page 314

WAREMA Renkhoff SE
Marktheidenfeld, Germany
www.warema.de
page 314

Weber Maschinenbau GmbH
Breidenbach, Germany
www.weberweb.com
page 314

Weber-Stephen
Deutschland GmbH
Ingelheim, Germany
page 314

Weingut Adams
Ingelheim am Rhein,
Germany
page 315

weinor GmbH & Co. KG
Köln, Germany
www.weinor.de
page 315

Wein & Vinos GmbH
Berlin, Germany
page 315

WEIREI INTERNATIONAL
Co., Ltd.
Taipei, Taiwan
www.twhhc.com
page 315

Wesemann GmbH
Syke, Germany
www.wesemann.com
page 316

WEST CORPORATION
Neyagawa-City, Osaka,
Japan
www.west-lock.co.jp
page 316

WE & WIN CONSTRUCTION
Taipei City, Taiwan
page 316

Whirlpool (China)
Investment Co., Ltd.
Shanghai, China
page 316

Wieland Electric GmbH
Bamberg, Germany
www.wieland-electric.de
page 316

Wiha Werkzeuge GmbH
Schonach, Germany
www.wiha.com
page 316

WILA Lichttechnik
Iserlohn, Germany
www.wila.com
page 317

Wilkhahn
Bad Münder, Germany
www.wilkhahn.de
page 317

will Magazine Verlag GmbH
München, Germany
www.novumnet.de
page 317

WILO SE
Dortmund, Germany
www.wilo.com
page 317

Wingman Condoms B. V.
Delft, Netherlands
www.wingmancondoms.com
page 317

Winora-Staiger GmbH
Sennfeld, Germany
www.winora-group.de
page 318

winwood48edition.com
Freiburg, Germany
www.winwood48edition.com
page 318

Withings
Issy Les Moulineaux, France
www.withings.com
page 319

W.L. Gore & Associates GmbH
Feldkirchen-Westerham,
Germany
www.gorebikewear.com
page 319

wodtke GmbH
Tübingen, Germany
www.wodtke.com
page 319

Wrack- und
Fischereimuseum Cuxhaven
Cuxhaven, Germany
page 319

WUHU MIDEA KITCHEN
AND BATH
Wuhu, China
page 319, 320

WUXI
little swan company limited
Wuxi, China
www.littleswan.com
page 320

X

XAL GmbH
Graz, Austria
www.xal.com
page 320, 321

X-BIONIC®
Asola, Italy
page 321, 322

Xiamen Solex
Technology Co., Ltd.
Xiamen, China
www.solex.com.cn
page 322, 323

Xindao
Rijswijk Zh, Netherlands
page 323

Xing Fu Cheng
New Taipei City, Taiwan
page 323

X-TEC GmbH
St. Margarethen an der
Raab, Austria
www.xtec.at
page 323

Y

Yakima Products, Inc.
Beaverton, OR,
United States of America
www.yakima.com
page 324

Yamaha Corporation
Hamamatsu, Japan
page 324

YAMAHA MOTOR CO., Ltd.
Shizuoka, Japan
www.yamaha-motor.co.jp
page 324

Yauchi Hamono
Sakai, Japan
www.yauchi-hamono.com
page 324

YG Entertainment
Seoul, South Korea
page 325

Yishouyuan
Bienenprodukte GmbH
Beijing, China
page 325

YNOT – HK (YMO)
Hong Kong
page 325

INDEX Manufacturer/Client

Yoga
Münster, Germany
www.tinaschmincke-yoga.de
page 325

YUN-YIH DESIGN COMPANY
Taipei, Taiwan
www.yundyih.com.tw
page 325

Zweibrüder Optoelectronics
Solingen, Germany
www.zweibrueder.com
page 328, 329

Z

Zaunkönig Lederhandwerk
Düsseldorf, Germany
www.zaunkoenig-
lederhandwerk.com
page 128

ZDF
Mainz, Germany
page 92

Zehnder Group Boleslawiec
Sp. z o. o
Boleslawiec, Poland
page 325

Zeichen & Wunder GmbH
München, Germany
www.zeichenundwunder.de
page 326

Zeutschel GmbH
Tübingen, Germany
www.zeutschel.de
page 326

Zhejiang Dahua Technology
Co., Ltd.
Hangzhou, China
page 326

Zhejiang Medzone Medical
Devices
Shaoxing, China
page 326

Zound Industries
Stockholm, Sweden
page 327

ZTE Corporation
Shanghai, China
www.zte.com.cn
page 327

Zumtobel AG
Dornbirn, Austria
www.zumtobelgroup.com
page 327

Zumtobel Lighting GmbH
Dornbirn, Austria
www.zumtobel.com
page 327, 328

369

INDEX

DESIGNERS

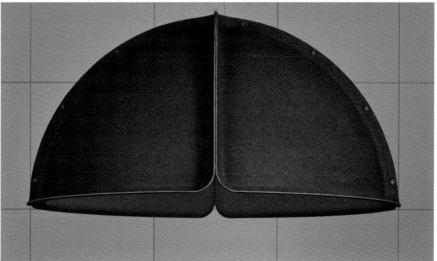

INDEX Design

Symbole

1&1 Internet AG
Karlsruhe, Germany
page 46

2nd West
Rapperswil, Switzerland
page 105

3deluxe
Wiesbaden, Germany
www.3deluxe.de
page 291

3pc GmbH Neue
Kommunikation
Berlin, Germany
page 219

3st digital GmbH
Mainz, Germany
www.3st-digital.de
page 311

3st kommunikation GmbH
Mainz, Germany
www.3st.de
page 96, 177, 291, 311

13&9 Design GmbH
Graz, Austria
www.13and9design.com
page 309

20PLUS
Seoul, South Korea
page 262

21TORR GmbH
Reutlingen, Germany
www.21torr.com
page 297

[d]tom
Fürth, Germany
www.d-tom.com
page 316

A

Aalto + Aalto
Helsinki, Finland
page 163

A&B One Digital GmbH
Berlin, Germany
www.a-b-one-digital.de
page 87

Acer Inc.
New Taipei City, Taiwan
page 47

ACME Europe
Kaunas, Lithuania
www.acme.eu
page 47

acme life
industrial design Co., Ltd.
Shenzhen, China
page 290

Acrylic couture
Remagen-Kripp, Germany
www.acrylic-couture.com
page 48

Activision
Santa Monica, CA,
United States of America
www.djhero.com
page 278

act&react
Werbeagentur GmbH
Dortmund, Germany
page 297

Adamidesign
München, Germany
www.adamidesign.de
page 68

Adam Opel AG
Rüsselsheim, Germany
page 48

Adlens Ltd.
Oxford, United Kingdom
www.adlens.com
page 48

admembers advertising
GmbH
Duisburg, Germany
www.admembers.de
page 48

Aekyung Industry
Seoul, South Korea
page 49

A & E OPTICAL Ltd.
Taipei, Taiwan
www.iplustec.com
page 49

Agentur am Flughafen
Altenrhein, Switzerland
page 49, 50

Agentur romen
Emmerich am Rhein,
Germany
www.romen-agentur.de
page 235

Agenzia di Comunicazione
Lugano, Switzerland
page 234

a.g Licht
Bonn, Germany
page 61

AhnLab
Seongnam-si, Gyeonggi-do,
South Korea
page 50

AIPHONE CO., Ltd.
Nagoya, Japan
www.aiphone.com
page 50

AiQ Smart Clothing Inc.
Taipei City, Taiwan
www.aiqsmartclothing.com
page 50

AIRACE ENTERPRISE Co., Ltd.
Tali Dist., Taichung City,
Taiwan
www.airace.com.tw
page 50, 51

Albers Inc.
New Taipei City, Taiwan
www.albers-creation.com
page 51

albert ebenbichler
innovations
Esslingen, Germany
www.albert-ebenbichler.com
page 53

Albert Leuchten
Fröndenberg, Germany
www.gebr-albert.de
page 51

Alfred Kärcher GmbH & Co. KG
Winnenden, Germany
page 52

ALPEN-MAYKESTAG GmbH
Puch, Austria
www.alpenmaykestag.com
page 54

ALTER
MONTROUGE PARIS, France
www.etaplighting.com
page 119

Alt Group
Auckland, New Zealand
www.altgroup.net
page 125, 213, 238

amp – audible brand and
corporate communication
München, Germany
page 194

André Konrad
Basel, Switzerland
page 174

Andreas Ringelhan
Industrialdesign
München, Germany
page 221

Andy Tong Interiors Ltd.
Hong Kong
www.atccl.com
page 325

Anish Kapoor Studio
London, United Kingdom
www.anishkapoor.com
page 327

Antec Mobile Products
(A. M. P.)
Fremont, CA,
United States of America
www.antec.com
page 56

antwerpes AG
Köln, Germany
www.antwerpes.de
page 110

Apple
Cupertino, CA,
United States of America
page 56, 57

Aran Research &
Development
Caesaeea Business Park,
Israel
www.aran-rd.com
page 150, 206

Arbonia AG
Arbon, Switzerland
www.arbonia.ch
page 57

Arcadyan Technology
Corporation
Hsinchu, Taiwan
www.arcadyan.com
page 57

ARCA-SWISS
Besançon, France
page 57

Arçelik A. S.
Istanbul, Turkey
page 58, 59

archipool.de
Karlsruhe, Germany
www.archipool.de
page 211

Arch. Matteo Astolfi
Treviso, Italy
www.matasto.com
page 215

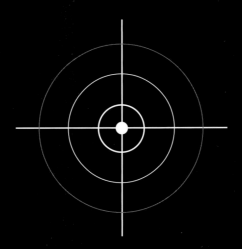

HITTING THE MARK
Nothing else !

INSTEAD OF A SHOT IN THE DARK!

AGENTUR FÜR MARKEN**TRÄUME**

WWW.A-M-T.NET

Arco Contemporary
Furniture
Winterswijk, Netherlands
www.arco.nl
page 59

Ardagh MP Germany GmbH
Erftstadt, Germany
www.ardaghgroup.com
page 59

Arda (Zhe Jiang)
Electric Co., Ltd.
Zhe Jiang Yong Kang, China
www.ardaappliance.com
page 59

Argent Alu
Kruishoutem, Belgium
www.argentalu.com
page 60

Arkoslight
Ribarroja del Turia, Spain
www.arkoslight.com
page 60

Arman Design and
Development
Istanbul, Turkey
www.armantasarim.com
page 119

Armor Manufacturing Corp.
Taipei, Taiwan
www.cole-products.com
page 60

Armstrong DLW GmbH
Bietigheim-Bissingen,
Germany
www.armstrong.eu
page 60

ARMZ Inc.
Tokyo, Japan
www.armz.co.jp
page 216

ARTEFAKT product design
Darmstadt, Germany
www.artefakt.de
page 159, 169

Artemide S. p. A.
Pregnana Milanese, Italy
www.artemide.com
page 61

Artlinco A / S
Horsens, Denmark
www.artlinco.com
page 104

ARUP
London, United Kingdom
www.arup.com
page 328

Ashcraft Design
Northridge, CA,
United States of America
page 144

ASUS Design Center
Taipei, Taiwan
page 62 – 64

at-design GbR
Fürth, Germany
www.atdesign.de
page 272, 273

ATELIER ALLURE
by THOMAS HAUSER
Wien, Austria
page 64, 65

ATELIER BRÜCKNER
Stuttgart, Germany
www.atelier-brueckner.de
page 49, 71, 212

ATELIER OPA
Tokyo, Japan
www.atelier-opa.com
page 302

atelier peter schmidt,
belliero & zandée
Hamburg, Germany
page 69

atelier schneeweiss GmbH
Schmieheim, Germany
www.atelier-schneeweiss.com
page 244

Atsuhito Kitora
Tokyo, Japan
page 239

AUDI AG
Ingolstadt, Germany
www.audi.de
page 65, 66

AUG design Co., Ltd.
Hangzhou, China
page 103, 326

Augenstein Produktdesign
Neu-Ulm, Germany
page 76

AVerMedia
TECHNOLOGIES, Inc.
New Taipei City, Taiwan
page 68

AVEXIR Technologies
Corporation
Zhubei City, Taiwan
www.avexir.com
page 68

AWOL Company
Calabasas, CA,
United States of America
page 305

B

BALMUDA Inc.
Tokyo, Japan
page 69

Bang & Olufsen
Struer, Denmark
www.bang-olufsen.com
page 70

Bartenbach GmbH
Aldrans / Tirol, Austria
www.bartenbach.com
page 328

basalte
Merelbeke, Belgium
www.basalte.be
page 70

base.io
Aarhus, Denmark
https://base.io
page 67

Bassier, Bergmann & Kindler
Ludwigsburg, Germany
www.bb-k.com
page 111

Beijing FromD Design
Consultancy
Beijing, China
www.fromd.net
page 161, 162

Bel Epok GmbH
Köln, Germany
page 71

Benedito Design
San Cugat Del Vallés,
Barcelona, Spain
page 122

BenQ Corp.
Taipei, Taiwan
page 72

Berker GmbH & Co. KG
Schalksmühle, Germany
page 72

Berlinerblau GmbH
Berlin, Germany
page 218

Bertelsmann SE & Co. KGaA
Gütersloh, Germany
www.bertelsmann.de
page 72

betec Licht AG
Dachau, Germany
page 73

BETO ENG. & MKTG. Co., Ltd.
Taichung, Taiwan
www.aplus-beto.com.tw
page 73

bfs·batterie füllungs
systeme GmbH
Bergkirchen bei München,
Germany
page 73

Biegert & Funk Product
GmbH & Co. KG
Schwäbisch Gmünd,
Germany
www.qlocktwo.com
page 73

bittermann industriedesign
Gelnhausen, Germany
www.bittermann-design.de
page 262

BlackBerry
Industrial Design Team
Waterloo, Canada
www.blackberry.com
page 74

Black Diamond
Equipment Limited
Salt Lake City, UT,
United States of America
www.
blackdiamondequipment.com
page 74

Blackmagic
Industrial Design Team
Melbourne, Australia
www.blackmagic-design.com
page 75

BLANCO GmbH & Co. KG
Oberderdingen, Germany
www.blanco-germany.com
page 75

Blankdesign
Weißenhorn, Germany
page 52, 53

Bleijh Industrial Design
Amsterdam, Netherlands
www.sandwichbikes.com
page 75

blm Filmproduktion GmbH
Hamburg, Germany
www.blm-film.de
page 67

COLE® The Art of Innovation.

COLE www.cole-products.com

DSA2®
Dynamic Spoke Alignment
Dynamic Spoke Alignment II

Cole ® High Compression Carbon Fiber

Bluebird
Seoul, South Korea
page 76

BMA Ergonomics
Zwolle, Netherlands
page 77

BMW AG
München, Germany
page 77, 78

BMW Group
München, Germany
www.bmwgroup.de
page 78 – 80

BMW Group
DesignworksUSA
München, Germany
page 62, 77, 264

BMW Group
DesignworksUSA
Newbury Park, CA,
United States of America
www.designworksusa.com
page 204, 275

Bodum Design Group
Triengen, Switzerland
page 80

Böhler Corporate Industrial Design
Fürth, Germany
page 154

BÖHLER
Corporate Industrial Design
Fürth, Germany
www.boehler-design.de
page 52

Böhler GmbH
Fürth, Germany
www.boehler-design.de
page 154

Bold°_a design company
Rio de Janeiro, Brazil
www.bolddesign.com.br
page 200

Bolin Webb Ltd.
Uppingham, United Kingdom
www.bolinwebb.com
page 81

Bombardier Transportation
Hennigsdorf, Germany
page 81

Bould Design
Mountain View, CA,
United States of America
www.bould.com
page 156

BRANDIS Industrial Design
Nürnberg, Germany
www.brandis-design.de
page 243, 266, 273

Braun Design
Kronberg i. T., Germany
www.braun.com
page 82

Braun Design | Household
Neu-Isenburg, Germany
page 104

BRAUNWAGNER GmbH
Aachen, Germany
www.braunwagner.de
page 82, 230

BRAVAT Plumbing Industrial Co., Ltd.
GuangZhou, China
page 83

BREE Designteam
Isernhagen, Germany
www.bree.com
page 83

Brighten the Corners Ltd.
London, United Kingdom
www.brightenthecorners.com
page 327

Britax Römer
Kindersicherheit GmbH
Ulm, Germany
www.britax-roemer.de
page 84

Brother Industries, Ltd.
Nagoya, Japan
page 84, 85

Bsize Inc.
Odawara, Kanagawa, Japan
www.bsize.com
page 87

Budde Industrie Design GmbH
Münster, Germany
www.budde-design.de
page 95, 96, 176, 185

BUFFALO Inc.
Nagoya, Japan
page 87

Buffon Wang
Taipei, Taiwan
page 271

Büro4 AG
Zürich, Switzerland
www.buero4.ch
page 277

Büro Destruct
Bern, Switzerland
www.burodestruct.net
page 67

büroecco
Kommunikationsdesign GmbH
Augsburg, Germany
www.bueroecco.com
page 182

busk + hertzog
London, United Kingdom
www.busk-hertzog.com
page 128

Busse
Design+Engineering GmbH
Elchingen, Germany
www.busse-design.com
page 309

BUWON ELECTRONICS Co., Ltd.
Daegu, South Korea
page 88

C

CaderaDesign
Würzburg, Germany
www.caderadesign.de
page 304, 314

Calibre Style Ltd.
Taipei, Taiwan
page 88

Canon Inc.
Tokyo, Japan
page 89, 90

Carlo Colombo Architect
Milano, Italy
www.carlocolombo.it
page 295

Carlos Martinez
Architekten AG
Berneck, Switzerland
www.carlosmartinez.ch
page 238

CARREFOUR
Les Ulis, France
www.carrefour.com
page 90, 91

Casa Rex
São Paulo, Brazil
www.casarex.com
page 64

C. Creative Inc.
Tokyo, Japan
www.c-creative.net
page 222

CCRZ sa
Balerna, Switzerland
www.ccrz.ch
page 234

cd's associates
Seoul, South Korea
page 236

cdupartners
Seoul, South Korea
www.cdunion.co.kr
page 198

Cebien
Gwangju-si, Gyeonggi-do, South Korea
page 91

Celesio AG
Stuttgart, Germany
www.celesio.com
page 91

CE Lighting Ltd.
Shenzhen, China
www.landlite.com
page 91

CELLULAR GmbH
Hamburg, Germany
www.cellular.de
page 91, 92

Cerebrum Design Co., Ltd.
Pathumthani, Thailand
www.studiocerebrum.com
page 211

Cesare Carlesso – Designer
Bassano del Grappa, Italy
page 194

Cheil Germany GmbH
Düsseldorf, Germany
page 104

ChenFangXiao Design International
Xiamen, China
www.cdi.hk
page 93

CHEN YU-MING
Kaohsiung, Taiwan
page 93

ChoisTechnology Co., Ltd.
Incheon, South Korea
wwww.x-pointer.com
page 93

INDEX Design

Chragokyberneticks
Bern, Switzerland
www.chragokyberneticks.ch
page 67

Christoph Behling Design Ltd.
London, United Kingdom
www.christophbehlingdesign.com
page 131

Christopher Redfern
Milano, Italy
page 328

Chung Choon
Seoul, South Korea
www.chung-choon.kr
page 50

Chungho Nais Co., Ltd.
Bucheon, Kyunggi-Do, South Korea
page 93

Chung Hua University
Hsinchu, Taiwan
www.shkinetic.com
page 94

Chung Yuan Christian University
Taoyuan, Taiwan
page 138

Cimax Design Engineering (Hong Kong) Ltd.
Shenzhen, China
www.libodesign.com
page 94

Cisco
Lysaker, Norway
www.cisco.com
page 94

CJ CheilJedang
Seoul, South Korea
page 94, 95

CLAUS KOCH™
Hamburg, Germany
www.clauskoch.com
page 110

Clever Pack
Rio de Janeiro, Brazil
www.cleverpack.com.br
page 96

Clormann Design GmbH
Penzing, Germany
www.clormanndesign.de
page 96, 140, 230, 317

COBO DESIGN CO., Ltd.
Nagoya, Japan
www.cobodesign.co.jp
page 50

code2design
Stuttgart, Germany
www.code2design.de
page 146, 222, 272

COMMANDANTE BERLIN GmbH
Berlin, Germany
www.commandante.org
page 311

COMMAX
Kyunggi-Do, South Korea
www.commax.com
page 96

Communication Design Laboratory
Tokyo, Japan
www.cdlab.jp
page 166, 293

Compal Experience Design
Taipei City, Taiwan
www.compal.com
page 97, 98

Compass pools Europe
Senec, Slovakia
page 98

comunicAzione
Rancate, Switzerland
page 234

CONARY ENTERPRISE CO., Ltd.
Taipei, Taiwan
page 99

Concertino
Tokyo, Japan
page 171

CONECTO Business Communication GmbH
Kaprun, Austria
www.conecto.at
page 54

Continental Reifen Deutschland GmbH
Hannover, Germany
page 99

Continuum
West Newton, PA, United States of America
www.continuuminnovation.com
page 204

Coordination Ausstellungs GmbH
Berlin, Germany
www.coordination-berlin.de
page 99

Cordivari Design
Morro D'Oro – TE, Italy
www.cordivaridesign.it
page 100

Coreana Cosmetics Co., Ltd.
Gyeonggi-do, South Korea
www.coreana.co.kr
page 100

COUGAR
Tainan, Taiwan
page 100

COWAY
Seoul, South Korea
page 100, 101

COWON SYSTEMS Inc.
Seoul, South Korea
www.cowon.com
page 101

CRE8 DESIGN
Taipei, Taiwan
www.cre8-designstudio.com
page 100

Creative Company KKI
Seoul, South Korea
page 91

CROSSPOINT
Seoul, South Korea
page 170

Crown Gabelstapler GmbH & Co. KG
München, Germany
www.crown.com
page 101

Cube Design China
Shenzhen, China
www.cubechina.net
page 270

Curious About
Zürich, Switzerland
page 277

cyclos-design GmbH
Münster, Germany
www.cyclos-design.de
page 210

CYPHICS Co., Ltd.
Seoul, South Korea
www.cyphics.com
page 157

D

DAEHAN A&C
Seoul, South Korea
www.daehananc.com
page 102

Daelim Dobidos
Incheon, South Korea
www.dlt.co.kr
page 102

Daimler AG
Stuttgart, Germany
www.daimler.com
page 102, 103

Dallmer GmbH & Co. KG
Arnsberg, Germany
www.dallmer.de
page 103

Daniela Fortinho Zilinsky
Blumenau, Brazil
page 103

DAPU tech. Co., Ltd.
Hangzhou, China
page 103

DART
Stuttgart, Germany
www.dartwork.de
page 103

D'ART DESIGN GRUPPE GmbH
Neuss, Germany
www.d-art-design.de
page 264

Dauphin Design-Team
Hersbruck, Germany
page 88

David Chipperfield Architects Ltd.
London, United Kingdom
www.davidchipperfield.com
page 327

Daylight Design Inc.
San Francisco, CA, United States of America
www.daylightdesign.com
page 196

DDB Tribal Wien
Wien, Austria
page 104, 203

Deff Corporation
Osaka City, Japan
deff.co.jp
page 104

defortec GmbH
Tübingen / Dettenhausen, Germany
www.defortec.de
page 113, 122, 192, 226

DELTA LIGHT N. V.
Wevelgem, Belgium
page 105

INDEX Design

DENSO CORPORATION
Aichi, Japan
www.globaldenso.com
page 106

DEPRO INTERNATIONAL
ASSOCIATES
Osaka, Japan
www.deprodesign.com
page 146

Design 3
Hamburg, Germany
www.design3.de
page 155, 156, 196

design Adrian
und Greiser GbR
Rheda-Wiedenbrück,
Germany
design-ag.de
page 71, 307

designaffairs GmbH
München, Germany
www.designaffairs.com
page 320

Design Ballendat Germany
Simbach am Inn, Germany
www.ballendat.com
page 209

Design Board Brussels
Brussels, Belgium
www.designboard.com
page 106

design büro groiss peter
Laakirchen, Austria
www.dbgp.at
page 266

DESIGN CENTER
MARCOPOLO
Caxias do Sul, Brazil
page 201

Designfactor
Kiel, Germany
www.designfactor.de
page 238

design Frank Greiser
Rheda-Wiedenbrück,
Germany
www.design-ag.de
page 165

Design Fuze Co., Ltd.
Seoul, South Korea
www.designfuze.com
page 93

Designhaus p&m
Berlin, Germany
www.designhaus-pm.de
page 293, 294

DesignIT
Copenhagen NV, Denmark
page 134

Designit Munich GmbH
München, Germany
www.designit.com
page 111, 133

Designliga –
Büro für Visuelle
Kommunikation und
Innenarchitektur
München, Germany
www.designliga.com
page 67

Design Partners
Bray, Co Wicklow, Ireland
www.designpartners.com
page 46, 88, 197

design-people
Aarhus C, Denmark
www.design-people.dk
page 215, 309

DESIGN S
Daegu, South Korea
www.designs.or.kr
page 209

designship
Ulm, Germany
www.designship.de
page 231

DesignSOHO
Seoul, South Korea
page 276

Design Studio S
Tokyo, Japan
www.design-ss.com
page 223

Design Tech
Ammerbuch, Germany
www.designtech.eu
page 296

design workroom brill
Giheung-gu, Yongin-si,
Gyeonggi-do, South Korea
www.brill.co.kr
page 93

Deutsche Telekom AG
Bonn, Germany
www.telekom.com
page 107, 108

DGT (DORELL.GHOTMEH.
TANE ARCHITECTS)
Paris, France
www.dgtarchitects.com
page 94

D&I
Brügge, Germany
page 225

Dialogform GmbH
Taufkirchen, Germany
www.dialogform.de
page 177

DI – Design Industrial Ltda.
São Paulo, Brazil
www.designindustrial.com.br
page 317

DIE LÖSUNG –
Strategie & Grafik
Memmingen, Germany
www.veraendertdiewelt.de
page 108

DIEPHAUS Betonwerk GmbH
Vechta, Germany
www.diephaus.de
page 108

digitalDigm
Seoul, South Korea
www.d2.co.kr
page 109

Digitalwerk GmbH
Wien, Austria
www.digitalwerk.at
page 177

DigitasLBi AG
Köln, Germany
www.digitaslbi.de
page 62

DIJIYA PATGEAR
Taipei, Taiwan
www.patgears.com
page 109

dingfest I design
Erkrath, Germany
www.dingfest.de
page 313

dittlidesign GmbH
Luzern, Switzerland
www.dittlidesign.ch
page 54

D-Link Corporation
Taipei, Taiwan
page 109

D.O.E.S
Seoul, South Korea
page 110

Do Ho Suh
London, United Kingdom
page 159

DOMY FACTORY
Seoul, South Korea
www.vimeo.com/domy
page 215

Dongdao Design Co., Ltd.
Beijing, China
www.dongdao.net
page 110, 200, 266, 298, 325

d'ORIGIN
Seoul, South Korea
page 170

DORMA
Ennepetal, Germany
www.dorma.com
page 111

Dräger Safety AG & Co.
KGaA
Lübeck, Germany
www.draeger.com
page 111

Drawing and Manual Inc.
Tokyo, Japan
www.drawingandmanual.jp
page 129, 168

d'strict
Seoul, South Korea
www.dstrict.com
page 175, 261

DURAN Group's
technical team
Mainz, Germany
page 112

Duscholux AG
Thun, Switzerland
www.duscholux.ch
page 113

Dyson
LONDON, United Kingdom
page 114

E

Eastern Global Corporation
New Taipei City, Taiwan
www.eg.com.tw
page 114

Eckstein Design
München, Germany
www.eckstein-design.com
page 134

Eco-Products Inc.
Boulder, CO,
United States of America
page 114

Mr. Viktor Bilak, Certified Designer

Member of the Designers' Unions of Moscow and Russia.
The prize-winner of the 18th International competition for product design concepts BraunPrize2012.
The prize-winner of the National Design Contest Russian Victory – 2010.
Since 2003 the head of EXPOLEVEL design-bureau.

The goal of the Slavyanskiye Oboi Company exhibition stand is to present wall paper, the main company product, in some abstract way. The core element of the stand is the wave-looking pylon that combines the images of a tree and an unfolding roll of wall-paper. The wave-looking pylons have been gathered in groups to result in the walls of complicated irregular shape, "dancing walls", creating quite distinctive image of a forest. The dominant of the stand color concept is the natural color of a tree trunk, so the tints of beige and the textured deep brown are used.

Project
Slavyanskiye Oboi
Design
Viktor Bilak
EXPOLEVEL
Moscow, Russia
www.expolevel.ru
Client
Slavyanskiye Oboi
Location
MosBuild-2012
Moscow, Russia

ECO Schulte GmbH & Co. KG
Menden, Germany
www.eco-schulte.de
page 114

Eczacibasi Yapi Gerecleri
Istanbul, Turkey
page 114, 115

Edifier Technology Co., Ltd.
Dongguan, China
www.edifier-international.com
page 115

E. I. Corporation
Hanam, South Korea
www.motherscornworld.com
page 115, 116

Eidenbenz
Industrial Design Est.
Triesen, Liechtenstein
page 73

eins:33 GmbH
München, Germany
www.einszu33.de
page 82

elbkind GmbH
Hamburg, Germany
www.elb-kind.de
page 117

ELECOM Co., Ltd.
Chuouku, Osaka, Japan
www.elecom.co.jp
page 116

Electrolux Group Design
Stockholm, Sweden
www.electrolux.com
page 116

eliumstudio
Paris, France
page 205

eliumstudio
Paris, France
www.eliumstudio.com
page 294

EliumStudio
Paris, France
www.eliumstudio.com
page 319

ellenbergerdesign
Bremen, Germany
www.ellenbergerdesign.de
page 55

ELR Co., Limited
Selangor, Malaysia
www.elr-group.com
page 117

ELSETA
Vilnius, Lithuania
www.elseta.com
page 48

Elsner Elektronik GmbH
Gechingen, Germany
www.elsner-elektronik.de
page 117

Elvin Textile Company
Bursa, Turkey
www.elvinfabrics.com
page 117

E-Make Tools Co., Ltd.
Taichung, Taiwan
page 118

EMAMIDESIGN
Berlin, Germany
www.emamidesign.de
page 299

emart
Seoul, South Korea
page 118

Emart Company Ltd.
Seoul, South Korea
page 118

ENSALT
Seoul, South Korea
www.ensalt.com
page 302, 303

Enthoven Associates
Antwerpen, Belgium
www.ea-dc.com
page 307

EOOS Design GmbH
Wien, Austria
www.eoos.at
page 112

EOOS Design GmbH
Wien, Austria
www.eoos.com
page 314

ERCO GmbH
Lüdenscheid, Germany
www.erco.com
page 119

Erdmann Design
Brugg, Switzerland
www.erdmann.ch
page 289

erfi – Ernst Fischer
GmbH & Co. KG
Freudenstadt, Germany
www.erfi.de
page 119

ERGO-FORM design
Colbitz, Germany
www.ergo-form.de
page 55

estúdio lógos
São paulo, Brazil
www.estudiologos.com.br
page 69

Eva Solo A / S
Måløv, Denmark
www.evasolo.com
page 120

eventlabs GmbH
Hamburg, Germany
page 222

Existence Design Co., Ltd.
Taichung, Taiwan
page 236

Eysing Design Kiel
Kiel, Germany
page 166

F

Fabian Bremer
Berlin, Germany
page 214

FabLab Shibuya
Tokyo, Japan
page 168

Fabrique
Delft, Netherlands
www.fabrique.nl
page 113

facts and fiction GmbH
Köln, Germany
www.factsfiction.com
page 121

Falcon White
Castrop-Rauxel, Germany
www.falconwhite.de
page 290

FAMILY BUSINESS
Stockholm, Sweden
page 59

Fantian Brand Management
Consultant Co., Ltd.
Shenyang, China
www.fantian.net
page 268

FARM
Rio de Janeiro, Brazil
www.farmrio.com.br
page 122

Fashion group Hyung ji
Seoul, South Korea
www.hyungji.co.kr
page 121

Favour Light Company
Limited
San Po Kong, Hong Kong
www.favourlight.com
page 122

FAXSON Co., Ltd.
Taichung City, Taiwan
page 122

Feish Design Co., Ltd.
Hangzhou, China
www.feish.com.cn
page 320

Festo AG & Co. KG
Esslingen, Germany
page 123, 124

fg branddesign
Stuttgart, Germany
www.fg-branddesign.com
page 108

Fibar Group
Poznan, Poland
www.fibaro.com
page 124

Fissler GmbH
Idar-Oberstein, Germany
www.Fissler.de
page 125

Fitbit Inc.
San Francisco, CA,
United States of America
www.fitbit.com
page 125

FLEX/the
INNOVATIONLAB B. V.
Delft, Netherlands
www.flex.nl
page 160, 275

FLEYE
Hedehusene, Denmark
www.fleye.dk
page 126

flora&faunavisions GmbH
Berlin, Germany
page 222

FLOWAIR Sp.j.
Gdynia, Poland
www.flowair.com
page 126

Flöz Industrie Design
Essen, Germany
www.floez.de
page 76

FLUVIA CONCEPT, S. L. U.
Madrid, Spain
www.fluvia.com
page 126

Food Industry Research
and Development Institute
Hsinchu, Taiwan
www.firdi.org.tw
page 127

Ford Design | Ford Werke
GmbH
Köln, Germany
www.ford.de
page 128

formfreun.de
gestaltungsgesellschaft
Berlin, Germany
www.formfreun.de
page 230, 231

Formfusion
Köln, Germany
www.formfusion.net
page 180

Formherr Industriedesign
Braunschweig, Germany
page 111

Form Us With Love
Stockholm, Sweden
www.formuswithlove.se
page 65, 127

Foshan Nanhai Chevan
Kayl, Luxembourg
www.vanguardworld.com
page 128

f/p design GmbH
München, Germany
www.fp-design-gmbh.com
page 171, 179

fpm
München, Germany
www.factor-product.com
page 305

Freeimage Design
Taipei City, Taiwan
www.freeimage.com
page 292

Friemel Design
Darmstadt, Germany
page 219

frog
San Francisco, CA,
United States of America
page 128

Frost Produkt
Oslo, Norway
page 94

Fuenfwerken Design AG
Wiesbaden, Germany
page 68, 295, 314, 315

FUJIFILM Corporation
Minato-ku, Tokyo, Japan
www.fujifilm.com
page 128, 129

FUJITSU DESIGN Ltd.
Kawasaki, Japan
jp.fujitsu.com
page 130

function Technology AS
Bergen, Norway
page 130

FUN FACTORY GmbH
Bremen, Germany
page 130

G

Gaggenau Hausgeräte
GmbH
München, Germany
www.gaggenau.com
page 130

Gajah International Pte. Ltd.
Singapore
www.gajah.com.sg
page 220

Gavin Harris
Sydney, Australia
www.futurespace.com.au
page 244

GB Europe
Utrecht, Netherlands
page 135

GBO DESIGN
Helmond, Netherlands
www.gbo.eu
page 121

GD Midea Air-Conditioning
Equipment Co., Ltd.
FoShan, China
page 131

Gearlab Co., Ltd.
Taipei, Taiwan
page 115

GemVax & Kael
Bundang-gu, Seongnam-si,
Gyeonggi-do, South Korea
page 132

GENERATION DESIGN GmbH
Wuppertal, Germany
page 175

George P Johnson
Torrance, CA,
United States of America
www.gpj.com
page 216

Gessaga Hindermann GmbH
Zürich, Switzerland
www.gessaga-hindermann.ch
page 288

GIGANTEX
Changhua County, Taiwan
page 133

Gigaset
Communications GmbH
München, Germany
www.gigaset.com
page 133

Giotto's Ind. Inc.
New Taipei City, Taiwan
www.giottos.com
page 133

GIXIA GROUP Co.
Taipei, Taiwan
page 133, 134, 170, 305, 315

GK Dynamics Inc.
Tokyo, Japan
page 324

Global Consumer Design,
Whirlpool
Shanghai, China
page 316

Gneiss Group
Copenhagen K., Denmark
page 55

GOGANG Aluminium Co., Ltd.
Seoul, South Korea
page 134

GÖKSEL ARAS – OFİSLINE
Sivas, Turkey
www.ofisline.com.tr
page 135

Golden Sun
Home Products Ltd.
Hong Kong, China
www.goldensun.com.hk
page 135

Golucci International Design
Taipei, Taiwan
www.golucci.com
page 292

Google
Mountain View, CA,
United States of America
page 150

Google, Inc.
Mountain View, CA,
United States of America
page 135

Goth Design
Seoul, South Korea
page 93

GP designpartners GmbH
Wien, Austria
www.gp.co.at
page 62

Grass Jelly Studio
Taipei City, Taiwan
grassjelly.tv
page 106

Greco Design
Belo Horizonte, Brazil
www.grecodesign.com.br
page 286

Griffwerk GmbH
Blaustein, Germany
www.griffwerk.de
page 136

grintsch communications
GmbH & Co. KG
Köln, Germany
www.grintsch.com
page 205

Grohe AG
Düsseldorf, Germany
www.grohe.com
page 136

G-Star Raw C.V.
Amsterdam, Netherlands
page 288

Guangdong Ganlanz
Microwave Oven and
Electrical Appliances
Manufacturing Co., Ltd.
Shunde, China
page 137

Guangdong Midea Kitchen
Guang Dong, China
page 137

Guangdong OPPO Mobile
Telecommunications Corp., Ltd.
Dongguan, China
www.oppo.com
page 137

Guangzhou Shiyuan
Electronic Co., Ltd.
Guangzhou, China
www.cvte.cn
page 138

Guangzhou
Times Property Group
Guangzhou, China
page 138

Gunitech Corp.
Qionglin Township, Hsinchu
County, Taiwan
www.gunitech.com
page 138

Günter Hermann
Architekten
Stuttgart, Germany
page 49

GWA Bathrooms & Kitchens
Sydney, Australia
www.gwagroup.com.au
page 139

G-WISE DESIGN Co., Ltd.
Hsinchu, Taiwan
page 271

H

Hadi Teherani AG
Hamburg, Germany
www.haditeherani.com
page 230

Haibike Design Team
Sennfeld, Germany
page 318

Haier
Innovation Design Center
Qingdao, China
page 139, 140

Hair O'right
International Corp.
Taoyuan County, Taiwan
page 140

Hakuhodo
Tokyo, Japan
page 180, 216

Hancomm
Seoul, South Korea
www.hancomm.co.kr
page 302, 303

Hangulhwa
Seoul, South Korea
page 141

Hangzhou Fan design Co., Ltd.
Hangzhou, China
www.weibo.com
page 306

Hangzhou ROBAM
Appliances Co., Ltd.
Hangzhou, Zhejiang
Province, China
www.robam.com
page 141

Hangzhou Teague
Technology Co., Ltd.
HangZhou, China
www.teague.net.cn
page 141

Hankook Tire Co., Ltd.
Seoul, South Korea
www.hankooktire.com
page 142

HAPPY IN COMPANY
Seoul, South Korea
www.sysmax.co.kr
page 143

HARIO Co., Ltd.
Tokyo, Japan
www.hario.jp
page 143

Harman Design Center
Shenzhen, China
page 143 – 145

Harman International
Industries, Inc.
Northridge, CA,
United States of America
www.harman.com
page 145

Harry Thaler
London, United Kingdom
www.harrythaler.it
page 215

hartmannvonsiebenthal
GmbH
Berlin, Germany
www.
hartmannvonsiebenthal.de
page 295

hartmut s. engel
design studio
Ludwigsburg, Germany
page 303

Hauser Lacour
Frankfurt a. M., Germany
www.hauserlacour.de
page 54, 89, 128, 201, 211, 263

Havas Worldwide Düsseldorf
Düsseldorf, Germany
www.havasworldwide.de
page 124

H.-D. SCHUNK GmbH & Co.
Mengen, Germany
www.schunk.com
page 147

Heine/Lenz/Zizka
Projekte GmbH
Berlin, Germany
www.hlz.de
page 163

Heine/Lenz/Zizka
Projekte GmbH
Frankfurt a. M., Germany
www.hlz.de
page 147, 164

Heine Warnecke
Design GmbH
Hannover, Germany
www.heinewarnecke.com
page 52

Hello AG
München, Germany
www.hello-muenchen.de
page 147

Hengyang 1more Electronic
Technology Co., Ltd.
Shenzhen, China
page 147

hers design Inc.
Toyonaka, Japan
www.hers.co.jp
page 146

Hettich ONI GmbH & Co. KG
Vlotho-Exter, Germany
www.hettich.com
page 148

Hewlett-Packard
Houston, TX,
United States of America
page 149, 150

Hewlett-Packard
San Diego, CA,
United States of America
page 149

Hewlett-Packard
Sunnyvale, CA,
United States of America
www.hp.com
page 149, 150

Heye GmbH
München, Germany
page 203

HGB
Hamburg, Germany
page 107

h&h design GmbH
Hagen, Germany
page 316

Hilti Corporation
Schaan, Liechtenstein
www.hilti.com
page 151, 152

Hisense
Industrial Design Center
Qingdao, China
www.hisense.com
page 152

Hitachi Koki Co., Ltd.
Hitachinaka-City, Ibaraki
Pref., Japan
www.hitachi-koki.co.jp
page 152, 153

Hitachi, Ltd.
Tokyo, Japan
www.hitachi.com
page 152

HMI Project
Karlstadt, Germany
www.hmi-project.com
page 314

Hochschule für Gestaltung
Offenbach a. M., Germany
page 61

HOFFMANN UND CAMPE
VERLAG GmbH
Hamburg, Germany
www.hocacp.de
page 70, 130

Honeywell Safety Products
Smithfield, RI,
United States of America
www.honeywellsafety.com
page 154

HORIBA, Ltd.
Kyoto, Japan
www.horiba.com
page 154

www.artize.in

CRAFTSMANSHIP IN WATER

Artize is a luxury bath brand from the Jaquar group. It has been introduced to cater to customers who aspire for luxury in their bath spaces. The brand aims at excellence at all levels and seeks to surpass global standards of design and quality. Artize products also adhere to all green building norms making them ideal for green projects.

Artize pays tribute to the fine traditions of exquisite craftsmanship and precision that have been the central focus of its manufacturing process since inception. Artize products are the result of a superior combination of human talent and machine.

*The brand promise for Artize is **"craftsmanship in water".** This phrase denotes the attention to detail, precision and design that converge to create the artize masterpieces.*

iF product design award 2014

Artize
craftsmanship in water

FAUCETS | SHOWERS | CERAMICS | WELLNESS

CRESCENT

INDEX Design

Hosoya Schaefer Architects AG
Zürich, Switzerland
www.hosoyaschaefer.com
page 67

HOZMI DESIGN
Kobe-City, Hyogo, Japan
www.hozmidesign.com
page 141

HSK Duschkabinenbau KG
Olsberg, Germany
page 155

HTC
New Taipei City, Taiwan
page 155

Huawei Device Co., Ltd.
Shenzhen, China
www.consumer.huawei.com
page 155, 156

Huawei Technologies Co., Ltd.
Shanghai, China
page 156

Huawei Technologies
Japan K. K.
Tokyo, Japan
www.huaweidevice.jp
page 155

Hultafors Group AB
Hultafors, Sweden
www.hultaforsgroup.com
page 156

Human Interface Design
Hamburg, Germany
www.human-interface.de
page 130

HUND B. communication
München, Germany
www.hundb.com
page 52

HUNTER DOUGLAS
EUROPE B. V.
Rotterdam, Netherlands
www.hunterdouglasgroup.com
page 157

Hurom L. S. Co., Ltd.
Gimhae-si, South Korea
www.hurom.com
page 157

hw.design GmbH
München, Germany
page 106, 158, 194, 213, 272

HYT Design Co., Ltd.
Tokyo, Japan
www.hyt-design.com
page 272

Hyundai Amco
Seoul, South Korea
page 158

Hyundai Card
Seoul, South Korea
page 158

Hyundai Design Center
Hwaseong, South Korea
page 158

HYVE Innovation
Design GmbH
München, Germany
page 126

I

icontrols
Sung-Nam-Si, Gyung-Gi-Do,
South Korea
www.icontrols.co.kr
page 159

IDEA DO IT
Daegu, South Korea
www.ideadoit.com
page 159

Idee und Klang
Basel, Switzerland
www.ideeundklang.com
page 67

identis, design-gruppe
joseph pölzelbauer
Freiburg, Germany
www.identis.de
page 286, 318

ID+IM Design Lab., KAIST
Daejeon, South Korea
www.idim.kaist.ac.kr
page 160

ID Infinity Limited
Hong Kong
page 160

IDL corporation
Tokyo, Japan
www.idlab.jp
page 221

ifm electronic
Essen, Germany
www.ifm.com
page 160

igus® GmbH
Köln, Germany
www.igus.de
page 160, 161

i/i/d Institut
für Integriertes Design
Bremen, Germany
www.iidbremen.de
page 206

ilusyd
Yongin, South Korea
page 163

Imagebakery
Seoul, South Korea
imagebakery.tv
page 159

Imago Design GmbH
Gilching, Germany
page 99

IMBUS design
Brno, Czech Republic
www.imbusdesign.cz
page 279

INDEED
Hamburg, Germany
www.indeed-innovation.com
page 173

Indeed Innovation GmbH
Hamburg, Germany
page 204

INDGROUP
Daegu, South Korea
page 164

Indio da Costa AUDT
Rio de Janeiro, Brazil
www.indiodacosta.com
page 224

Industrial Design Associates
Hechendorf, Germany
page 73

Industrial Design Center
of Haier Group
Qingdao, China
www.haier.com
page 139

Industrial facility
London, United Kingdom
www.industrialfacility.co.uk
page 148

industrialpartners GmbH
Beerfelden, Germany
www.industrialpartners.de
page 157

INFINITE
Seoul, South Korea
www.infinite.co.kr
page 211

INNOCEAN Worldwide
Seoul, South Korea
page 159

INNODesign, Inc.
Gyeung gi do, South Korea
www.innodesign.com
page 164

INSPIRE.D
Seoul, South Korea
www.inspired.crevisse.com
page 302, 303

Institute of Zhejiang
Medzone
Hangzhou, China
page 326

Intel Corporation
Shanghai, China
www.intel.com
page 117

Interbrand
Köln, Germany
www.interbrand.com
page 107

Interbrand Korea
Seoul, South Korea
page 165

intergram
Seoul, South Korea
www.intergram.co.kr
page 165

Involution Studios
Arlington, VA,
United States of America
www.goinvo.com
page 302

Ippolito Fleitz Group –
Identity Architects
Stuttgart, Germany
www.ifgroup.org
page 60, 71, 142, 285

IRIVER
Seoul, South Korea
www.iriver.com
page 167

iRobot Corporation
Bedford, PY,
United States of America
www.irobot.com
page 167

item Industrietechnik GmbH
Solingen, Germany
www.item24.com
page 168

LP Circle

Design: Mikkel Beedholm/KHR Arkitekter A/S. **LP CIRCLE** is a large product family of LED-based fixtures to suit all types of rooms and architecture. The fixtures come in diameters of 260 and 450 mm. They can be fully or partially recessed, surface-mounted or suspended. They are available in white, blue, red and black. Special light technology results in comfortable, energy efficient lighting. Several wattages and lumen packages are available, and a choice of 3000K or 4000K.

louis poulsen

www.louispoulsen.com

INDEX Design

ITRI Green Energy
& Environment
Hsinchu, Taiwan
www.itri.org.tw/eng
page 164

IXI Co., Ltd.
Tokyo, Japan
www.ixi-jp.com
page 94, 290

J

JaderAlmeida design &
arquitetura
Princesa, Brazil
page 278

Jakubowski – Büro für
Gestaltung
Ratingen, Germany
www.ralfjakubowski.com
page 170

James Irvine S. R. L.
Milano, Italy
page 252

jangled nerves
Stuttgart, Germany
www.janglednerves.com
page 67

Jaquar & Company Pvt. Ltd.
Gurgaon, Haryana, India
artize.in
page 169

Jawbone Industrial Co., Ltd.
New Taipei City, Taiwan
www.jawbone.com.tw
page 170

jehs + laub GbR
Stuttgart, Germany
page 67

jehs + laub GbR
Stuttgart, Germany
www.jehs-laub.com
page 85, 317

Jeju Special Self-governing
Province Development Corp.
Jeju, South Korea
www.jpdc.co.kr
page 170

Jens Kirchner
Düsseldorf, Germany
page 135

JIA Inc.
Hong Kong
page 171

Jiyoun Kim Studio
Seoul, South Korea
www.jiyounkim.com
page 76

JOBY Inc.
San Francisco, CA,
United States of America
www.joby.com
page 171

JOMOO Kitchen & Bath
Appliances Co., Ltd.
Xiamen, China
www.jomoo.com.cn
page 171, 172

Joseph Joseph
London, United Kingdom
www.josephjoseph.com
page 172

JoyaTech
Xiamen, China
page 172

Joy Industrial Co., Ltd.
Taichung Hsien, Taiwan
page 173

JPDC
Jeju, South Korea
page 170

JUNO
Hamburg, Germany
www.juno-hamburg.com
page 177, 232, 319

JUSTIME
Team of Shengtai Brassware
Chang Hua, Taiwan
www.justime.com
page 268

Just Mobile Ltd.
Taichung, Taiwan
www.just-mobile.com
page 173

K

Kähler Design A / S
Næstved, Denmark
www.kahlerdesign.com
page 174

Kai Stania Product Design
Wien, Austria
www.kaistania.com
page 321

Kan & Lau Design
Consultants
Kowloon Tong, Hong Kong
page 174

Karen Olze
Berlin, Germany
page 67

Kastl Design
München, Germany
www.kastldesign.de
page 62

Katharina Andes
Basel, Switzerland
page 174

KD Navien Co., Ltd.
Seoul, South Korea
page 175

Keim Identity GmbH
Zürich, Switzerland
www.keimidentity.ch
page 121

Kemper
Kommunikation GmbH
Frankfurt a. M., Germany
www.keko.de
page 112

KENJIRO NII AND DESIGN
Tokyo, Japan
page 239

Kenwood Limited
Hampshire, United Kingdom
page 175

Kesseböhmer GmbH
Bad Essen, Germany
www.kesseboehmer.de
page 175

Kia Design Team
Frankfurt a. M., Germany
www.kiamotors.com
page 159, 176

KIA Motors Corporation
Seoul, South Korea
www.kiamotors.com
page 158

KIDU
Daejeon, South Korea
www.kidu.co.kr
page 176

KIENLEDESIGN
Leinzell, Germany
www.kienledesign.de
page 173

KIMS HOLDINGS
Busan, South Korea
page 176

King & Miranda Design S. R. L.
Milan, Italy
page 325

Kinney Chan and Associates
Hong Kong
www.kca.com.hk
page 154

KircherBurkhardt GmbH
Berlin, Germany
www.kircher-burkhardt.com
page 53, 306, 177, 311

Kiska GmbH
Anif-Salzburg, Austria
www.kiska.com
page 204

Kjaerulff Design
Hadsund, Denmark
www.kjaerulff-design.dk
page 174

KLANGERFINDER
Stuttgart, Germany
page 71

Klaus Nolting . ON3D
Hamburg, Germany
www.on3d.de
page 166

kleiner und bold GmbH
Berlin, Germany
www.kleinerundbold.com
page 68, 106

KMS TEAM
München, Germany
www.kms-team.com
page 177, 211

KMW
Hwaseong-si, Gyeonggi-do,
South Korea
page 178

Koga B. V.
Heerenveen, Netherlands
www.koga.com
page 178

Kohler
China Investment Co., Ltd.
Shanghai, China
www.kohler.com.cn
page 178

Kohler Mira Ltd.
Cheltenham, United
Kingdom
page 178

Kokes Partners
Praha 9 Vysocany
Czech Republic
www.kokespartners.cz
page 279

Kolle Rebbe GmbH
Hamburg, Germany
www.kolle-rebbe.de
page 141

KOLON
Gwacheon-si, Gyeonggi-Do,
South Korea
page 179

KONE Design Solutions
Hyvinkää, Finland
www.kone.com
page 179

Konings & Kappelhoff
design agency
Amsterdam, Netherlands
www.koningskappelhoff.com
page 56

Koninklijke Gazelle N. V.
Dieren, Netherlands
page 180

KONO Design und
Technologie GmbH
Hannover, Germany
www.kono.de
page 284

Konrad Hornschuch AG
Weißbach, Germany
www.hornschuch.de
page 180

Konstantin Grcic
Industrial Design
München, Germany
www.konstantin-grcic.com
page 118

KONVERDI GmbH
Lehrte, Germany
www.konverdi.de
page 288

Koop Industrial Design
Hamburg, Germany
www.koopdesign.de
page 112

Koroyd
Monte-Carlo, Monaco
www.koroyd.com
page 277

KozSusani Design
Chicago, IL,
United States of America
www.kozsusanidesign.com
page 253

KREO CO., Ltd.
Tokyo, Japan
page 179

Krones AG
Neutraubling, Germany
page 181

kt media hub
Seoul, South Korea
page 181

Kucher & Thusbass
Amerang, Germany
www.kucherthusbass.de
page 318

Kuhn Rikon AG
Rikon, Switzerland
www.kuhnrikon.ch
page 181

KUKA AG
Augsburg, Germany
page 181

Kumho Tire Co.
Yongin-si, Gyeonggi-do,
South Korea
page 182

Kuoyenting
Taichung City, Taiwan
page 182

Kurz Kurz Design
Solingen, Germany
www.kurz-kurz-design.de
page 287

KW43 BRANDDESIGN
Düsseldorf, Germany
www.kw43.de
page 132, 136, 184, 209, 252, 297, 303, 304

KYE SYSTEMS Corp.
Taipei Hsien, Taiwan
page 183

KYOCERA Corporation
Yokohama-shi, Kanagawa,
Japan
page 183

kyowon L & C
Seoul, South Korea
page 183

L

labor b designbüro
Dortmund, Germany
www.laborb.de
page 146, 307

LABOR WELTENBAU
Stuttgart, Germany
www.laborweltenbau.de
page 83

LACO
Uhrenmanufaktur GmbH
Pforzheim, Germany
www.laco.de
page 184

LAMOTO
Köln, Germany
www.lamotodesign.com
page 165

Landon Yoon
Seoul, South Korea
www.landonyoon.com
page 302, 303

Landor and Associates
Hamburg, Germany
www.landor.com
page 184

Lattke und Lattke GmbH
Reichenberg, Germany
www.lattkeundlattke.de
page 311

LDE Belzner Holmes
Stuttgart, Germany
page 49, 71

Leadtek Research Inc.
Taipei, Taiwan
page 184

Leagas Delaney
Hamburg GmbH
Hamburg, Germany
www.leagasdelaney.de
page 53, 54

LED Linear GmbH
Neukirchen-Vluyn, Germany
www.led-linear.de
page 184

Lee Young Joo
Anyang, South Korea
page 168

LEICHT Küchen AG
Waldstetten, Germany
www.leicht.de
page 185

Leifheit AG
Nassau / Lahn, Germany
www.soehnle.de
page 185

Lengyel Design
Essen, Germany
www.lengyel.de
page 132

Lenovo
Morrisville, NC,
United States of America
page 185, 186

Lenovo (Beijing) Ltd.
Beijing, China
page 186

Leo Burnett GmbH
Frankfurt a. M., Germany
www.leoburnett.de
page 55, 124, 201, 216, 219

Lepel & Lepel
Köln, Germany
www.lepel-lepel.de
page 135

Lesmo GmbH & Co. KG
Düsseldorf, Germany
page 187

LessingvonKlenze
München, Germany
page 203

Lesmandesign
Greiz, Germany
page 291

leyess
Seoul, South Korea
www.lyanature.com
page 187, 200

LFI Photographie GmbH
Hamburg, Germany
page 187

LG Electronics Inc.
Seoul, South Korea
www.lg.com
page 187 – 192

LG Hausys
Seoul, South Korea
www.lghausys.com
page 192

LG innotek
Ansan, South Korea
www.lginnotek.com
page 192

Libratone A/S
Herlev, Denmark
www.libratone.com
page 192

Lichtwerke GmbH
Köln, Germany
www.lichtwerke.com
page 285

Liebherr
Kirchdorf / Iller, Germany
www.liebherr.com
page 192

INDEX Design

Lieblingsagentur GmbH
Krefeld, Germany
www.lieblingsagentur.de
page 48

LIFETHINGS
Seoul, South Korea
www.lifethings.in
page 302, 303

Lifetrons Switzerland AG
Dicken, Switzerland
www.lifetrons.ch
page 193

LIGHTperMETER
Waregem, Belgium
www.lightpermeter.be
page 193

Lightweight Containers
Den Helder, Netherlands
www.keykeg.com
page 193

Linde
Material Handling GmbH
Aschaffenburg, Germany
www.linde-mh.de
page 194

LINE
Seongnam-si, Gyeonggi-do,
South Korea
www.line.me
page 194

LI-NING (China) Sports
Goods Co.,
Beijing, China
page 194

linshaobin design
Shantou, China
www.linshaobin.com
page 195

Liquid Agency
San Jose, CA,
United States of America
page 135

Lite-on Technology Corp.
Taipei, Taiwan
page 168

LIULIGONGFANG
Shanghai, China
www.liuliliving.com
page 195

LOBTEX Co., Ltd.
Osaka, Japan
www.lobtex.co.jp
page 195

loew d.sign*
Klein-Winternheim,
Germany
www.loew-design.de
page 81

LOGICDATA GmbH
Deutschlandsberg, Austria
www.logicdata.at
page 196

Lord Benex
International Co., Ltd.
Taichung City, Taiwan
www.lordbenex.com
page 197

Lothar Böhm Associates
Hamburg, Germany
www.lba-branding.com
page 238

LOTTE Confectionery
Seoul, South Korea
page 198

Louis Poulsen Lighting A / S
København K, Denmark
page 198

LSA International
Sunbury on Thames,
United Kingdom
page 198

LUFTZUG CO., Ltd.
Amsterdam, Netherlands
page 283

lumini
São Paulo, Brazil
www.lumini.com.br
page 198, 199

LUNAR
München, Germany
page 199

LUNAR
San Francisco, CA,
United States of America
page 199

Lytro, Inc.
Mountain View, CA,
United States of America
page 200

M

Madeblunt Ltd.
Auckland, New Zealand
www.bluntumbrellas.com
page 200

ma design GmbH & Co. KG
Kiel, Germany
page 46

Magnus Pettersen
London, United Kingdom
page 162

Maltani lighting
Seoul, South Korea
www.taewon.co.kr
page 201

Manifesto Architecture P. C.
New York, NY,
United States of America
www.mfarch.com
page 235

Mario Mazzer
architect | designer
Conegliano, Italy
www.mariomazzer.it
page 202

MARKUS BISCHOF
produktdesign
Nürnberg, Germany
www.markusbischof.de
page 262, 286

MARTINELLI LUCE S. p. A.
Lucca, Italy
www.martinelliluce.it
page 202

Martin et Karczinski
München, Germany
www.martinetkarczinski.de
page 51, 66, 221, 313

Martin L. Daester
Windisch, Switzerland
page 174

Mathieu Lehanneur
since 1974 S. A. R. L
Paris, France
www.mathieulehanneur.fr
page 121

Matrix Audio Ltd.
ONT, Canada
www.matrixaudio.com
page 202

Matuschek Design &
Management
Aalen, Germany
www.matuschekdesign.de
page 151, 152

MAVIC S. A. S.
Annecy Cedex 9, France
page 202

MAXU
Technology European Office
Lyon, France
page 202

Medidea Corporation
Fukuoka, Japan
page 221

Medion AG
Essen, Germany
page 203

Medion Designteam
Essen, Germany
page 203

Mehnert Corporate Design
GmbH & Co. KG
Berlin, Germany
www.mehnertdesign.de
page 317

melting elements GmbH
Hamburg, Germany
page 288

Merit Co., Ltd.
Daejeon, South Korea
page 204

Merry Electronics Co., Ltd.
Taichung, Taiwan
page 205

MESO
Digital Interiors GmbH
Frankfurt a. M., Germany
www.meso.net
page 78

METAPHOR Inc.
Tokyo, Japan
www.metaphor.co.jp
page 216

Metrax GmbH
Rottweil, Germany
www.primedic.com
page 205

Metso Corporation
Helsinki, Finland
page 205

Metso Minerals, Inc.
Tampere, Finland
page 206

Microsoft
Redmond, WA,
United States of America
page 206

Miele & Cie. KG
Gütersloh, Germany
www.miele.de
page 207

LATERALO PLUS LED

Einzigartige Lichtqualität kombiniert mit zurückhaltender Transparenz

Unique light quality combined with subtle transparency

Design: Hartmut S. Engel

TRILUX
SIMPLIFY YOUR LIGHT.

MiGlaz Design Co., Ltd.
Nonthaburi, Thailand
www.miglaz.com
page 207

Milla & Partner
Innovationslabor
Stuttgart, Germany
page 207

MindDesign
London, United Kingdom
page 201

Mindsailors s.c. M.
Bonikowski R.Pilat
Poznan, Poland
www.mindsailors.com
page 131

m.i.r. media
Köln, Germany
www.mir.de
page 180

miscea GmbH
Berlin, Germany
www.miscea.com
page 207

Misfit Wearables
Corporation
Redwood City, CA,
United States of America
www.misfitwearables.com
page 207

MiTAC International Corp.
Taipei, Taiwan
page 208

Mit Biss.
Münster, Germany
www.mitbiss.com
page 325

MIT Media Lab
Cambridge, MA,
United States of America
page 92

Mitsubishi Electric
Corporation
Kamakura, Japan
page 208, 219

MITSUBISHI PENCIL Co., Ltd.
Tokyo, Japan
page 209

MIYAKE Design Office
Tokyo, Japan
page 253

Moldex-Metric AG & Co. KG
Walddorfhäslach, Germany
www.moldex-europe.com
page 209

MOMENT
Tokyo, Japan
www.moment-design.com
page 167, 168

MONK Design
Pontresina, Switzerland
www.monk-design.com
page 177

moo:u design
Seoul, South Korea
page 170

Morgen Digital
München, Germany
www.morgen-digital.com
page 195

Mosarte
Tijucas – SC, Brazil
www.mosarte.com.br
page 209

Motorola Solutions
Schaumburg, IL,
United States of America
www.motorolasolutions.com
page 210

mount inc.
Tokyo, Japan
www.mount.jp
page 180

Mu Creatives Co., Ltd.
Taipei, Taiwan
www.mu-creatives.com
page 134

muehlhausmoers
Berlin, Germany
www.muehlhausmoers.com
page 210

muehlhausmoers
Köln, Germany
www.muehlhausmoers.com
page 179

Multiple S. A. Global Design
La Chaux-de-Fonds,
Switzerland
www.multiple-design.ch
page 213

Murken Hansen Product
Design
Berlin, Germany
www.murkenhansen.de
page 233

Mutabor Design GmbH
Hamburg, Germany
www.mutabor.de
page 78

Myoung Ho Lee
Seoul, South Korea
page 302, 303

N

Nanjing HuaQi
Visual Design Co., Ltd.
Nanjing, China
www.huaqiart.com
page 211

Nanyang Technological
University
Singapore
page 93

NAOTO FUKASAWA DESIGN
Tokyo, Japan
page 61, 252

Native Design Ltd.
London, United Kingdom
page 88, 149, 299

NAVER Corp.
Gyeonggi-do, South Korea
page 212, 213

NEC Display Solutions Ltd.
Kanagawa, Japan
page 212

Neolab convergence
Seoul, South Korea
page 118, 213, 304

neo.studio
Berlin, Germany
www.neostudio.de
page 288

neplus
Seoul, South Korea
www.nep-plus.com
page 85

NewDealDesign LLC
San Francisco, CA,
United States of America
www.newdealdesign.com
page 125, 165

NEXENTIRE Corporation
Yangsan Kyungnam,
South Korea
www.nexentire.com
page 213

NHN Entertainment Corp
Seongnam-si, Gyeongg,
South Korea
page 213

Nico Bats
Frankfurt a. M., Germany
page 214

Nike, Inc.
Beaverton, OR,
United States of America
page 214

NIKE SPORTS KOREA CO., Ltd.
Seoul, South Korea
page 215

Nikon Corporation
Tokyo, Japan
page 215

NOA
Aachen, Germany
www.noa.de
page 142

Nokia
Beijing, China
page 217

Nokia
Espoo, Finland
www.nokia.com
page 216, 217

Nokia
London, United Kingdom
page 218

Nokia Corporation
Beijing, China
page 218

Nokia Corporation
Espoo, Finland
page 218

Nokia Design
Calabasas, CA,
United States of America
www.nokia.com
page 217

NONOBJECT
Palo Alto, CA,
United States of America
www.nonobject.com
page 196, 305

Nova Design (Shanghai) Ltd.
Shanghai, China
www.e-novadesign.com
page 210

NOVAREVO
Incheon, South Korea
www.novarevo.net
page 219

N+P Industrial Design GmbH
München, Germany
www.np-id.com
page 132

INDEX Design

npk design B. V.
Leiden, Netherlands
www.npk.nl
www.npkdesign.de
page 195, 216, 276

nr21 DESIGN GmbH
Berlin, Germany
www.nr21.com
page 266

NTT DOCOMO, Inc.
Tokyo, Japan
page 219

Nya Nordiska Textiles GmbH
Dannenberg, Germany
www.nya.com
page 220

O

Occhio GmbH
München, Germany
page 221

Office DO
Kyoto, Japan
page 154

O-I GLASSPACK
GmbH & Co. KG
Düsseldorf, Germany
page 304

Okamura Corporation
Tokyo, Japan
www.okamura.co.jp
page 221

Oktober
Kommunikationsdesign GmbH
Bochum, Germany
www.oktober.de
page 263

Olympus Corporation
Tokyo, Japan
www.olympus.com
page 222

OMRON HEALTHCARE Co., Ltd.
Kyoto, Japan
www.healthcare.omron.co.jp
page 222, 223

One Plus Partnership Ltd.
Hong Kong
page 156

One Plus Partnership Ltd.
Hong Kong
www.onepluspartnership.com
page 223

Onno Jetel e. K.
Gilching, Germany
page 201

OPTIX Digital Pictures GmbH
Hamburg, Germany
page 67

Ora ïto Mobility
Paris, France
page 224

ordinarypeople
Seoul, South Korea
www.ordinarypeople.kr
page 224

OREA Technology
(Beijing) Co., Ltd.
Beijing, China
www.oreadesign.com
page 164

Osram GmbH
München, Germany
www.osram.com
page 224

Ottenwälder und
Ottenwälder
Schwäbisch Gmünd,
Germany
www.ottenwaelder.de
page 153, 222

Otto Bock
Mobility Solutions GmbH
Königsee, Germany
www.ottobock.de
page 225

Otto Ganter GmbH & Co. KG
Furtwangen, Germany
www.ganter-griff.de
page 225

outermedia GmbH
Berlin, Germany
www.outermedia.de
page 289

Oval Design Ltd.
Kowloon, Hong Kong
www.ovaldesign.com.hk
page 89

Oventrop GmbH & Co. KG
Olsberg, Germany
www.oventrop.de
page 225

Oxylane
Lille, France
www.decathlon.com
page 225

Oxylane
Villeneuve d'Ascq, France
www.artengo.com
page 225

OZAKI INTERNATIONAL
Co., Ltd.
San Chung City, Taipei,
Taiwan
www.ozaki.us
page 226

Oz Design Ltda.
São Paulo, Brazil
www.ozdesign.com.br
page 226

P

Pacific Market International
Amsterdam, Netherlands
page 226

Panama
Werbeagentur GmbH
Stuttgart, Germany
www.panama.de
page 87

Panasonic Corporation
Hyogo, Japan
panasonic.co.jp
www.panasonic.de
page 227, 229

Panasonic Corporation
Osaka, Japan
page 229, 230

Panasonic Corporation
Tokyo, Japan
panasonic.co.jp/dc
page 227 – 229

PANTECH CO., Ltd.
Seoul, South Korea
page 230

Paul Hettich GmbH & Co. KG
Kirchlengern, Germany
www.hettich.com
page 148

PDR
Cardiff, United Kingdom
www.pdronline.co.uk
page 298

pearl creative
Ludwigsburg, Germany
www.pearlcreative.com
page 83, 243

Pearl Studios Inc.
Montreal QC, Canada
page 207

PearsonLloyd Design Ltd.
London, United Kingdom
www.pearsonlloyd.com
page 72

Pedro Paulo Venzon Filho
Erechim, Brazil
www.biccateca.com.br/volta
page 73

PEGACASA Design Team
Taipei, Taiwan
www.pegacasa.com
page 231

PEGA D&E
Taipei, Taiwan
www.pegadesign.com
page 46, 231

Pentagon Design Ltd.
Helsinki, Finland
page 235

pentree_calligraphy
Seoul, South Korea
www.pentree.net
page 93

Peter Schmidt Group
Hamburg, Germany
www.peter-schmidt-group.de
page 194

PF Concept International B. V.
Roelofarendsveen,
Netherlands
www.pfconcept.com
page 232, 233

Pfeffersack & Soehne
Koblenz, Germany
www.pfeffersackundsoehne.de
page 233

PFLITSCH GmbH & Co. KG
Hückeswagen, Germany
www.pflitsch.de
page 233

Phenix lighting
(Xiamen) Co., Ltd.
Xiamen, China
page 199

Philippe Starck
Paris, France
www.starck.com
page 113

Philips Design
Eindhoven, Netherlands
www.philips.com/design
page 244 – 252

INDEX Design

philosys Co., Ltd.
Gunsan-si, Jeollabuk-do,
South Korea
www.philosys.com
page 233

Phoenix Design
GmbH & Co. KG
Stuttgart, Germany
www.phoenixdesign.de
www.phoenixdesign.com
page 142, 143, 148, 166, 157,
308, 309

Phoenix Tapware
Bayswater, Australia
www.phoenixtapware.com.au
page 234

Pilotfish
Munich|Amsterdam|Taipei
Taipei City, Taiwan
page 169

Pipilotti Rist
Zürich, Switzerland
page 238

piranha grafik
Lichtenstein, Germany
page 289

PixelGrid Inc.
Tokyo, Japan
www.pxgrid.com
page 216

PIXOMONDO Studios
GmbH & Co. KG
Frankfurt a. M., Germany
www.pixomondo.com
page 78

Plantronics Design
Santa Cruz, CA,
United States of America
page 234, 235

Plants Associates Inc.
Toyama, Japan
www.arm-s.net/dseries
page 261

Plast Competence Center AG
Zofingen, Switzerland
www.plastcc.ch
page 262, 295

platinumdesign
Stuttgart, Germany
www.platinumdesign.com
page 133, 274

PLATOON KOMMUNIZIERT
Mainz, Germany
www.platoongraphics.com
page 177

Plus Eyewear Limited
Sanpokong, Hong Kong
www.pluseyewear.com
page 235

Pluspol Interactive
Leipzig, Germany
www.pluspol.info
page 117

PlusX
Seoul, South Korea
www.plus-ex.com
page 183, 275, 325

POLA Inc.
Tokyo, Japan
www.pola.co.jp
page 235

POLARWERK GmbH
Bremen, Germany
www.polarwerk.de
page 147

polyform Industrie Design
München, Germany
www.polyform-design.de
page 221

Porsche Design GmbH
Porsche Design Studio
Zell am See, Austria
www.porsche-design.com
page 99, 136

Porsche
Engineering Group GmbH
Weissach, Germany
page 194

Proad Identity
Taipei, Taiwan
www.proadidentity.com
page 169

ProDesign
Neu-Ulm, Germany
www.prodesign-ulm.de
page 88

Produkt DESIGN Wolf Heieck
Sontheim, Germany
www.design-heieck.de
page 52

Prof. Wulf Schneider
und Partner
Stuttgart, Germany
www.profwulfschneider.de
page 275

Projekttriangle
Design Studio
Stuttgart, Germany
www.projekttriangle.com
page 109

Promate Technologies
Shenzhen, China
www.promate.net
page 236

Puls Produktdesign
Darmstadt, Germany
www.puls-design.de
page 275

push design
Aachen, Germany
page 230

pxd, Inc.
Seoul, South Korea
www.pxd.co.kr
page 92

Q

q~bus Mediatektur GmbH
Berlin, Germany
www.q-bus.de
page 108

Qisda Creative Design Center
Taipei, Taiwan
www.qisda.com
page 67, 164, 236, 237

QNAP Systems, Inc.
New Taipei City, Taiwan
www.qnap.com
page 237

Quanta Computer Inc.
Tao Yuan, Taiwan
www.quantatw.com
page 237

QUECHUA
Domancy, France
www.quechua.com
page 237

Questto|Nó
São Paulo, Brazil
www.questtono.com
page 49, 263

R

Raum B Architektur
Zürich, Switzerland
www.raumb.ch
page 238

raumservice
Stuttgart, Germany
www.raumservice.de
page 110

RED
Stuttgart, Germany
www.redproducts.de
page 119

Redblackdesigns Inc.
Taipei, Taiwan
www.redblackdesigns.com
page 292

red pepper
Bremen, Germany
page 206

Reflect Inc.
Seoul, South Korea
page 239

Reflexion ag | Thomas Mika
Zürich, Switzerland
www.reflexion.ch
page 320

REFORM DESIGN Produkt
Stuttgart, Germany
www.reform-design.de
page 272

REINSCLASSEN GmbH & Co. KG
Hamburg, Germany
www.reinsclassen.de
page 146

Relations Inc.
Shibuya-ku, Tokyo, Japan
www.relationsgroup.co.jp
page 239

Renata Rubim
Porto Alegre – RS, Brazil
www.renatarubim.com.br
page 239

RenQing Technology Co., Ltd.
Shenzhen, China
www.ihave.hk
page 240

repaBAD GmbH
Wendlingen, Germany
page 240

RICOH
Yokohama-shi, Japan
www.ricoh.co.jp
page 240, 241

RINGEL & PARTNER
Trier, Germany
www.ringelundpartner.de
page 287

ripple effect Co., Ltd.
Nishinomiya, Japan
www.ripple-effect.org
page 324

SQUARO EDGE 12
PERFECT AESTHETICS

Squaro Edge 12 is made of Quaryl®, the exclusive material innovation from Villeroy & Boch. The distinguishing feature of Squaro Edge 12 is its slender 12mm rim. Experience special bathing comfort and a highly aesthetic design that harmonises beautifully with the flush-mounted outlet and overflow fittings with an optional, integrated water inlet available in chrome and white.

WWW.VILLEROY-BOCH.COM

INDEX Design

Robert Bosch
Hausgeräte GmbH
München, Germany
www.bosch-hausgeraete.com
www.bosch-home.com
page 85, 241 – 243

Robert Welch Designs Ltd.
Chipping Campden, United Kingdom
page 244

Rokitta Produkt &
Markenästhetik
Mülheim a. d. Ruhr, Germany
page 70

Ronald Scliar Sasson
Gramado, Brazil
page 267

Ruska, Martín,
Associates GmbH
Berlin, Germany
page 315

Ryohin Keikaku Co., Ltd.
Tokyo, Japan
www.muji.co.jp
page 253

Ryu Kozeki
Tokyo, Japan
www.ryukozeki.com
page 292

S

saad branding+design
Curitiba, Brazil
www.saad-studio.com
page 265

Saidi 'sign Büro für Produkt- und Grafikdesign
Zell u. A., Germany
page 288

Saint-Gobain Weber GmbH
Düsseldorf, Germany
www.sg-weber.de
page 253

SAKURA Inc.
Tokyo, Japan
www.sakura-tokyo.jp
page 179

Salvatore Tiles & Tapetes
Lindolfo Collor, Brazil
www.salvatoreminuano.com.br
page 253

Samkwang Glass Co., Ltd.
Seoul, South Korea
www.glasslock.co.kr
page 254

Samsung Electronics Co., Ltd.
Gyeonggi-do, South Korea
page 255

Samsung Electronics Co., Ltd.
Seoul, South Korea
www.samsung.com
page 254 – 260

Samsung SDS Co., Ltd.
Seoul, South Korea
page 261

Sankyo-Alumi Company,
Sankyo Tateyama Inc.
Toyama, Japan
page 261

SANTOX
Löffingen, Germany
www.santox.com
page 261

SANYO Electric Co., Ltd.
Hyogo, Japan
page 261, 262

SapientNitro
Köln, Germany
page 65, 106, 240

Sardi Innovation
Milano, Italy
www.sardi-innovation.com
page 300

Satoshi Yanagisawa
Tokyo, Japan
page 92

Sayuri Studio, Inc.
Tokyo, Japan
www.ss-studio.com
page 262

Scalae
Dalby, Sweden
www.scalae.se
page 134

Schaffner & Conzelmann AG
Basel, Switzerland
www.designersfactory.com
page 99, 178

Scherfdesign Concept & Development
Köln, Germany
www.scherfdesign.de
page 71

Scheufele Hesse Eigler
Kommunikation
Frankfurt, Germany
page 112

Schindelhauer Bikes
Berlin, Germany
www.schindelhauerbikes.de
page 263

Schirrmacher
Product Design
Landsberg, Germany
page 124

SCHMIDHUBER /
KMS BLACKSPACE
München, Germany
page 66, 82

Scholz & Volkmer GmbH
Wiesbaden, Germany
www.s-v.de
page 102, 263

schramke design
Berlin, Germany
www.schramke-design.de
page 120

Schuberth GmbH
Magdeburg, Germany
www.schuberth.com
page 264

Schüco International KG
Bielefeld, Germany
page 264

seepex GmbH
Bottrop, Germany
page 265

Seidldesign
Stuttgart, Germany
page 238, 239

Seiko Epson Corporation
Shiojiri-shi, Nagano-ken, Japan
page 265

SELF Electronics Co., Ltd.
Ningbo, China
www.self-electronics.com
page 265

Selic Industriedesign
Augsburg, Germany
www.selic.de
page 181

Sempio Foods Company
Seoul, South Korea
www.sempio.com
page 265

Seoul Women's Univ.
Seoul, South Korea
page 201

Sercomm Corporation
Taipei, Taiwan
www.sercomm.com
page 267

SERGE CORNELISSEN BVBA
Kortrijk (Marke), Belgium
www.sergecornelissen.com
page 166

SERIES NEMO, S. L.
Barcelona, Spain
www.seriesnemo.com
page 51

Shanghai United Imaging
Healthcare Co., Ltd.
Shanghai, China
page 267

Shang Yih
Interior Design Co., Ltd.
Taipei, Taiwan
page 267

Shan Shui
Branding Design Studio
Zhongli City, Taoyuan County, Taiwan
page 138

Shenzhen ARTOP Design Co., Ltd.
Shenzhen, China
www.artopcn.com
page 270

Shenzhen Baojia Battery
Tech Co., Ltd.
Shenzhen, China
www.mipow.com
page 268, 269

Shenzhen Breo Technology Co., Ltd.
Shenzhen, China
www.breocare.com
page 270

Shenzhen Daidea industrial
product design Co., Ltd.
Shenzhen, China
www.daidea.com.cn
page 270

Shenzhen Jiangyi Science
& Technology Development Co., Ltd.
Shenzhen, China
page 270

Shenzhen Rapoo
Technology Co., Ltd.
Shenzhen, China
page 270

INDEX Design

Shenzhen Uoshon
Communication Technology
Co., Ltd.
Shenzhen, China
www.uoshon.com
page 270

Shiftcontrol
Copenhagen, Denmark
page 67

Shigeru Ban Architects
Tokyo, Japan
www.shigerubanarchitects.com
page 127

Shilo
New York, NY / San Diego, CA,
United States of America
page 212

Shimano Inc.
Osaka, Japan
page 271

Shimano Inc.
Sakai City, Japan
www.shimano.com
page 271

SID Design
Seoul, South Korea
page 215

sieger design GmbH & Co. KG
Sassenberg, Germany
www.sieger-design.de
www.sieger-design.com
page 95, 113, 136

Siemens AG
Forchheim, Germany
page 272

Siemens AG, Industry Sector
Nürnberg, Germany
page 272, 273

Siemens
Electrogeräte GmbH
München, Germany
page 85, 86, 273, 274

SIG Combibloc
Linnich, Germany
www.sig.biz
page 274

Sign Architecture
& Interior Design Co., Ltd.
Taipei, Taiwan
signarchi.com.tw
page 274

Silicon Power Computer
& Communications Inc.
Taipei, Taiwan
www.silicon-power.com
page 274

SILO Lab.
Seoul, South Korea
www.silolab.kr
page 215

SIMIZ Technik
Suita-City, Osaka, Japan
www.simiz.co.jp
page 141

Simon & Goetz Design
GmbH & Co. KG
Frankfurt a. M., Germany
www.simongoetz.de
page 253

simple GmbH
Köln, Germany
www.simple.de
page 275

SKYLOTEC GmbH
Neuwied, Germany
www.skylotec.de
page 276

Smaller International Co., Ltd.
Taipei, Taiwan
page 276

S.M. Entertainment
Seoul, South Korea
page 276, 277

Smith Optics
Ketchum, ID,
United States of America
www.smithoptics.com
page 277

SNA Europe [Sweden] AB
Enköping, Sweden
www.bahco.com
page 277

Sodastream Ltd.
Ben Gurion Airport, Israel
page 278

SOFTBANK BB Corp.
Tokyo, Japan
page 202

SOLARLUX Aluminium
Systeme GmbH
Bissendorf, Germany
www.solarlux.de
page 278

Soma
San Francisco, CA,
United States of America
www.drinksoma.com
page 278

Sonnenstaub – Büro für
Gestaltung und Illustration
Berlin, Germany
www.sonnenstaub.com
page 46, 236

SonoScape Co., Ltd.
Shenzhen, China
page 279

Sonos Inc.
Santa Barbara, CA,
United States of America
page 279

SONY Corporation
Tokyo, Japan
www.sony.net/design
page 279 – 283

SONY Europe Ltd.
Surrey, United Kingdom
page 280 – 282

Sony Mobile
Communications Inc.
Tokyo, Japan
page 283, 284

Sony PCL Inc.
Shinagawa-ku, Tokyo, Japan
page 283

Spacetalk
Seoul, South Korea
www.webspacetalk.co.kr
page 158

Spark Design & Innovation
Rotterdam, Netherlands
www.sparkdesign.nl
page 193

speziell®
Offenbach, Germany
www.speziell.net
page 272

SPRGO Design
Xiamen, China
page 172

S. Siedle & Söhne
Furtwangen, Germany
www.siedle.de
page 285

Staatliches Bauamt
Traunstein
Traunstein, Germany
page 71

STABILO International GmbH
Nürnberg, Germany
page 285

Stadler Form
Aktiengesellschaft
Zug, Switzerland
page 285

STAEDTLER
Mars GmbH & Co. KG
Nürnberg, Germany
www.staedtler.de
page 286

STAHLWILLE
Wuppertal, Germany
www.stahlwille.de
page 286

Steelcase S. A.
Schiltigheim, France
www.steelcase.fr
page 287

SteelSeries ApS
Valby, Denmark
www.steelseries.com
page 287

Stefan Diez
München, Germany
www.stefan-diez.com
page 82

Stefan Radinger
Linz, Austria
www.stefanradinger.com
page 187

Stelton
Espergærde, Denmark
page 287

stephan gahlow
produktgestaltung
Hamburg, Germany
www.gahlow.de
page 198

stereolize GmbH
München, Germany
www.stereolize.com
page 100

STI Group
Lauterbach, Germany
www.sti-group.com
page 288

stories within architecture
Berlin, Germany
www.sw-architecture.de
page 184

INDEX Design

Strike Communications
Seoul, South Korea
www.strike-design.com
page 50

Studio Aisslinger
Berlin, Germany
page 126

STUDIO CHAPEAUX
Hamburg, Germany
www.studiochapeaux.com
page 83

Studio de Bevilacqua
Milano, Italy
page 61

Studio Klass
Milano, Italy
www.studioklass.com
page 127

studiomem
München, Germany
www.studiomem.de
page 125

Studio Oeding GmbH
Hamburg, Germany
page 76, 77, 92, 115

studioQ
Wien, Austria
www.studioq.at
page 117

Studio Qiao
Taipei City, Taiwan
page 195

Studio Sonda
Vizinada, Republic of Croatia
page 101

Sudhaus GmbH & Co. KG
Iserlohn, Germany
page 289

Suh Architects
Seoul, South Korea
www.suharchitects.com
page 159

Sunrise Medical
GmbH & Co. KG
Malsch, Germany
www.sunrisemedical.de
page 290

Suunto Design Team
and Typolar Oy
Vantaa, Finland
www.suunto.com
page 290

Swareflex GmbH
Vomp, Austria
www.swareflex.com
page 291

SYZYGY Deutschland GmbH
Frankfurt a. M., Germany
www.syzygy.de
page 112, 176

T

tado° GmbH
München, Germany
www.tado.com
page 291

Taiwan Tech
Taipei City, Taiwan
page 165

Takagi Co., Ltd.
Kitakyushu, Japan
page 292, 293

Talantone Creative Group
Seoul, South Korea
www.talantone.com
page 254

Talocci Design
Rome, Italy
www.taloccidesign.com
page 295

TAMSCHICK MEDIA+SPACE
Berlin, Germany
page 71

Tangerine & Partners
Daejeon, South Korea
www.tangerine.net
page 163

Tangible
Seoul, South Korea
www.tangiblebd.com
page 293

Tatic Designstudio S. R. L.
Milano, Italy
www.taticdesignstudio.com
page 111

TBWAHAKUHODO
HAKUHODO Inc.
Tokyo, Japan
page 239

TCS AG
Triberg, Germany
www.carus-concepts.com
page 293, 294

TEAGUE
Seattle, WA,
United States of America
www.teague.com
page 233

TEAM 7
Natürlich Wohnen GmbH
Ried im Innkreis, Austria
www.team7.at
page 294

teamandproducts GmbH
München, Germany
page 294

Teams Design
Chicago, IL,
United States of America
page 81

Teams Design GmbH
Esslingen, Germany
page 52, 81, 127

Teams Design
Shanghai, China
page 81, 82

Teataster Group
Holdings Limited
Taipei, Taiwan
page 294

Teatulia Organic Teas
Denver, CO,
United States of America
www.teatulia.com
page 294

Tesseraux + Partner
Potsdam, Germany
page 175, 176

Tetra Pak
Modena, Italy
www.tetrapak.com
page 295

TGS Design Consultancy
Shenzhen, China
page 231

The Brand Union GmbH
Hamburg, Germany
www.thebrandunion.de
page 296

The Brand Union Paris
Paris, France
page 125

The Hamptons Bay –
Design Company
München, Germany
www.thehamptonsbay.com
page 46, 74

therefore Ltd.
London, United Kingdom
www.therefore.com
page 299

THIMM
Verpackung GmbH & Co. KG
Northeim, Germany
www.thimm.de
page 296

Thinkware
Seongnam-si, South Korea
page 296

Thoma+Schekorr
Berlin, Germany
www.thoma-schekorr.com
page 107

Thomas Biswanger Design
Ingolstadt, Germany
www.thomasbiswanger.de
page 72

Thonet GmbH
Frankenberg (Eder),
Germany
www.thonet.de
page 297

Thule Group
Longmont, CO,
United States of America
www.thulegroup.com
page 297

Thule Group
Seymour, CT,
United States of America
www.thulegroup.com
page 297

ThyssenKrupp Uhde GmbH
Dortmund, Germany
www.thyssenkrupp-uhde.de
page 297

tisch13 GmbH
München, Germany
www.tisch13.com
page 66

Tobias Grau GmbH
Rellingen, Germany
www.tobias-grau.com
page 298

Tombow Pencil Co., Ltd.
Tokyo, Japan
www.tombow.com
page 298

Tom Leifer Design
Hamburg, Germany
www.tomleiferdesign.de
page 187

ela
The elastic hinge cover

DEFNE KOZ
MARCO SUSANI
design

soul of furniture

Tommee Tippee
Cramlington, United Kingdom
www.tommeetippee.com
page 299

Tools Design
Copenhagen, Denmark
www.toolsdesign.com
page 173

TopGun / Tforce / Ubike
Yuan Lin Town, Changhua, Taiwan
www.ubike-tech.com
page 299

Top Victory Electronics (Taiwan) Co., Ltd.
New Taipei City, Taiwan
page 299

TORAFU ARCHITECTS
Shinagawa-ku Tokyo, Japan
page 283

Toshiba
Research Triangle Park, NC, United States of America
page 300

Toshiba
Tokyo, Japan
www.toshiba.co.jp
page 300, 301

TOTO Ltd.
Tokyo, Japan
www.toto.co.jp
page 301, 302

Toyo Seikan Group Holdings, Ltd.
Tokyo, Japan
page 216

Toyota Material Handling Europe AB
Mjölby, Sweden
www.toyota-forklifts.eu
page 147

TP-LINK Design
Nanshan, Shenzhen, China
page 302

Tribecraft AG
Zürich, Switzerland
www.tribecraft.ch
page 132, 223

Tricon Design AG
Kirchentellinsfurt, Germany
www.tricon-design.de
page 71

Trippen A. Spieth, M. Oehler GmbH
Berlin, Germany
www.trippen.com
page 303

true fruits GmbH
Bonn, Germany
www.true-fruits.com
page 304

TSUBAKIMOTO Chain Company
Kyoto, Japan
www.tsubakimoto.com
page 304

Tt Design
Taipei, Taiwan
www.ttdesignworks.com
page 296

TU Dresden
Dresden, Germany
page 231

Tupperware General Services N. V.
Aalst, Belgium
www.tupperware.com
page 304

Twisthink
Holland, MI, United States of America
page 101

U

U:GO Designers Office
Nara, Japan
www.ugo-d.com
page 154

Uneka
Livermore, CA, United States of America
page 135

Unitron
Kitchener, Canada
www.unitron.com
page 305

Unity Studios
Aarhus, Denmark
www.unity-studios.com
page 67

Universal Everything
Sheffield, United Kingdom
www.universaleverything.com
page 159

Uponor Group
Vantaa, Finland
www.uponor.com
page 305

urbn; interaction
Berlin, Germany
www.urbn.de
page 306

ushowdesign Co., Ltd.
New Taipei City, Taiwan
www.ushowdesign.com
page 174

V

V2 Studios Ltd.
London, United Kingdom
page 188

Vaillant Inhouse Design
Remscheid, Germany
page 306, 307

VALD Design Co., Ltd.
Taipei, Taiwan
page 165

VanBerlo B. V.
Eindhoven, Netherlands
www.vanberlo.nl
page 120, 182, 238

Van den Weghe
Zulte, Belgium
www.lapris.be
page 307

Van Hoecke
Sint-Niklaas, Belgium
www.taor.com
page 307

VELUX A / S
Hørsholm, Denmark
www.velux.com
page 307

Verdom
Maceió, Brazil
page 308

Vermop Salmon GmbH
Gilching, Germany
www.vermop.com
page 308

Veryday
Bromma, Sweden
www.veryday.com
page 49, 183, 291

VESTEL
Manisa, Turkey
www.vestel.com.tr
page 308

Vetica Group
Luzern, Switzerland
www.vetica-group.com
page 293, 326

Villeroy & Boch AG
Mettlach, Germany
www.villeroy-boch.de
page 309

vitamin e GmbH
Hamburg, Germany
www.agenturlabor.com
page 222

VITE! Concepts GmbH
Frankfurt, Germany
www.viteconcepts.de
page 53

VitrA Karo
Istanbul, Turkey
www.vitra.com.tr
page 309, 310

Vogel's Products B. V.
Eindhoven, Netherlands
page 310

Völkl Sports GmbH & Co. KG
Straubing, Germany
www.voelkl.com
page 310

Volkswagen AG
Wolfsburg, Germany
www.volkswagen.de
page 310

Volkswagen Design
Wolfsburg, Germany
page 207

Volvo Construction Equipment
Gothenburg, Sweden
www.volvoce.com
page 312

Volvo Trucks Product Design
Gothenburg, Sweden
www.volvotrucks.com
page 312

V-ZUG AG
Zug, Switzerland
www.vzug.ch
page 312

W

Wacker Neuson Linz GmbH
Linz-Leonding, Austria
www.neusonkramer.com
page 96

INDEX Design

Wagner System GmbH
Lahr, Germany
page 313

Wang I-Hsuan Cindy
Singapore
www.cindy-wang.com
page 131

WAREMA Renkhoff SE
Marktheidenfeld, Germany
www.warema.de
page 314

we'd design
Seongnam-si, Gyeonggi-do,
South Korea
www.wed-design.com
page 292

weinor GmbH & Co. KG
Köln, Germany
page 315

weiss
communication + design
Biel, Switzerland
www.wcd.ch
page 284

weissraum.de(sign)°
Hamburg, Germany
www.weissraum.de
page 146

WeLL Design
Utrecht, Netherlands
www.welldesign.com
page 212

WEN SHENG LEE
ARCHITECTS & PLANNERS
Taipei City, Taiwan
page 323

Wen Sheng Lee
Architects & Planners
Taipei, Taiwan
page 316

Werbung etc.
Stuttgart, Germany
www.werbungetc.de
page 52

Werksdesign Volker
Schumann
Berlin, Germany
www.werksdesign.de
page 296

wesentlich. visuelle
kommunikation
Aachen, Germany
www.wesentlich.com
page 72

WEST CORPORATION
Neyagawa-City, Osaka,
Japan
www.west-lock.co.jp
page 316

Whipsaw Inc.
San Jose, CA,
United States of America
page 112, 120

White Elements GmbH
Berlin, Germany
www.whiteelements.com
page 48

whiteID Integrated Design
Schorndorf, Germany
www.white-id.com
page 99

WHITEvoid
interactive art & design
Berlin, Germany
www.whitevoid.com
page 159

Wiha Werkzeuge GmbH
Schonach, Germany
www.wiha.com
page 316

WILA Lichttechnik
Iserlohn, Germany
www.wila.com
page 317

Wingman Condoms B. V.
Delft, Netherlands
www.wingmancondoms.com
page 317

winkelbauer-design
Ludwigsburg, Germany
www.winkelbauer-design.de
page 234

Winora Engineering Team
Sennfeld, Germany
page 318

wirDesign
communications AG
Berlin, Germany
www.wirdesign.de
page 115

wirDesign
communications AG
Braunschweig, Germany
www.wirdesign.de
page 74, 285

W.L. Gore & Associates GmbH
Feldkirchen-Westerham,
Germany
www.gorebikewear.com
page 319

wodtke GmbH
Tübingen, Germany
www.wodtke.com
page 319

Wolf Production GmbH
Frechen-Königsdorf,
Germany
www.wolf-production.com
page 78

Work
Oslo, Norway
page 94

WOW Inc.
Shibuya-ku Tokyo, Japan
page 283

WUHU MIDEA
KITCHEN AND BATH
Shunde, China
page 319, 320

WUXI
little swan company limited
Wuxi, China
www.littleswan.com
page 320

X

XAL GmbH
Graz, Austria
www.xal.com
page 321

Xiamen Solex
Technology Co., Ltd.
Xiamen, China
www.solex.com.cn
page 322, 323

Xindao
Rijswijk Zh, Netherlands
page 323

X-TEC GmbH
St. Margarethen an der
Raab, Austria
www.xtec.at
page 323

X-Technology Swiss R & D AG
Wollerau, Switzerland
www.x-technology.com
page 321, 322

Y

Yakima Products, Inc.
Beaverton, OR,
United States of America
www.yakima.com
page 324

Yamaha Corporation
Hamamatsu, Japan
page 324

yellow design
Köln, Germany
www.yellowdesign.com
page 78

YS design Inc.
Osaka, Japan
page 223

YUAN YUAN STUDIO
Beijing, China
page 237

YUMAMAN CREATIVE &
DESIGN CO., Ltd.
Taipei, Taiwan
page 264

YUN-YIH DESIGN COMPANY
Taipei, Taiwan
www.yundyih.com.tw
page 325

Yves Behar – fuseproject
San Francisco, CA,
United States of America
www.fuseproject.com
page 66

Z

Zeichen & Wunder GmbH
München, Germany
www.zeichenundwunder.de
page 139, 164, 326

Zeug Design GmbH
Salzburg, Austria
www.zeug.at
page 252

ZhangYangsheng design
studio
Shanghai, China
www.okinok.com
page 87

Zhejiang Dahua Technology
Co., Ltd.
Hangzhou, China
page 326

ziba tokyo Co., Ltd.
Tokyo, Japan
page 239

ZIBERT + FRIENDS GmbH
München, Germany
www.zibert.com
page 311

zooom production GmbH
Fuschl am See, Austria
www.zooom.at
page 175

Zound Industries
Stockholm, Sweden
page 327

ZTE Corporation
Shanghai, China
www.zte.com.cn
page 327

Zum Kuckuck GmbH & Co. KG
Würzburg, Germany
www.zumkuckuck.com
page 204, 311, 312

Zumtobel Lighting GmbH
Dornbirn, Austria
www.zumtobel.com
page 328

Zweibrüder Optoelectronics
Solingen, Germany
www.zweibrueder.com
page 328, 329